LIFE IN A GALL

LIFE IN A GALL

THE BIOLOGY AND ECOLOGY OF INSECTS THAT LIVE IN PLANT GALLS

ROSALIND BLANCHE

CSIRO

PUBLISHING

National Library of Australia Cataloguing-in-Publication entry:

    Blanche, Rosalind.

    Life in a gall : the biology and ecology of insects that live in plant galls / by Rosalind Blanche.

    9780643106437 (pbk.)
    9780643106444 (epdf)
    9780643106451 (epub)

    Includes bibliographical references and index.

    Galls (Botany)
    Gall insects.
    Insect-plant relationships.
    632.2

Published by
CSIRO PUBLISHING
36 Gardiner Road, Clayton VIC 3168
Private Bag 10, Clayton South VIC 3169
Australia

Telephone: [+613] 9545 8555
Local call: 1300 788 000 (Australia only)
Fax: +61 3 9662 7555
Email: csiropublishing@csiro.au
Web site: www.publishing.csiro.au

Front cover: Larva, pupa and adult jewel beetle *Ethonion leai* on *Dillwynia hispida*.
Back cover: Stem galls of the jewel beetle *Ethonion leai* on *Dillwynia hispida* showing the holes where adults have emerged from their galls. Photos: Peter Lang.

Set in Minion Pro 9.5/11
Cover and text design by James Kelly
Typeset by Oryx Publishing Pty Ltd
Printed by Ingram Lightning Source

CSIRO PUBLISHING publishes and distributes scientific, technical and health science books and journals from Australia to a worldwide audience and conducts these activities autonomously from the research research activities of the Commonwealth Scientific and Industrial Research Organisation (CSIRO). The views expressed in this publication are those of the author and do not necessarily represent those of, and should not be attributed to, the publisher or CSIRO.

Feb26_RP_ILS

Contents

Preface and acknowledgements

Life in a Gall aims to reveal the interesting, and sometimes surprising, facts about Australian gall-inducing insects and the native plant species that host them. It explores the tantalising question of how insects induce plants to form galls and the significance of galls for the gall-inducing insects, their ecological communities and for humans. It also outlines the strategies employed by different insect groups to find a suitable site to induce a gall, adapt to living part of their lives in the confined spaces of a gall, obtain food, mate and escape the gall.

Gall-inducing insects have enemies that prey on them, and the book outlines the various ways in which they defend themselves from these enemies. It highlights the serious problems that gall-inducing insects can cause in home gardens and for horticulture, agriculture, forestry and conservation. It also explains why gall-inducing insects are sometimes able to build up to pest proportions and suggests ways in which outbreaks can be controlled.

But gall-inducing insects do have a positive side. They may benefit human society by providing free ecosystem services such as control of invasive weeds, pollination of figs and production of food.

A final chapter provides tips for people who want to collect and study galls for themselves. It also briefly outlines current scientific research associated with Australian gall-inducing insects – including an unusual research study carried out by primary school children with the help of some scientists. The success of this study demonstrates what can be achieved with a little collaboration!

I have intended *Life in a Gall* to be a reference guide for interested readers, so the use of scientific terms is kept to a minimum. Most new concepts are introduced using general language followed by the appropriate scientific term in brackets. If at any time you want to check the definition of a scientific word there is a glossary (page 62), which lists relevant scientific terms and their meanings.

I would particularly like to acknowledge the contribution made by the numerous photographers who so generously donated most of the images appearing in this book. No single person could have had the time, or knowledge, to photograph and verify such a diverse array of insect and plant species. Without the input of these photographers this book would never have happened.

In addition, I would like the many people who expressed enthusiasm for the idea of a book about Australian galls to know how extremely encouraging their comments were. I also want to thank my husband, John Ludwig, who, in his usual calm and unassuming way, provided emotional and technical support during the whole of the book's development. Thanks too to my helpful CSIRO editor, Nick Alexander, and to CSIRO Publishing for agreeing to produce the book.

Rosalind Blanche

Female galls of two different scale insect species of the genus *Apiomorpha* with two very different appearances – even though they are on the same eucalypt. Photo: Lyn Cook.

Introduction

At some time in their lives most people have probably noticed strange lumps, bumps or deformities on leaves, stems or flower buds of plants. They may well wonder what causes these unusual growths and how they might affect the plant. Few people realise that many of these deformations, or galls as they are called, are the result of an insect inducing the plant to provide it with a place to live and feed.

The term 'plant gall' refers to any abnormal swelling of plant tissue. The swelling may be due to an increase in the size of plant cells (hypertrophy) and/or an increase in the number of cells (hyperplasia). It is important to understand that the gall is formed by the plant in response to the presence of a gall-inducing agent. The gall-inducing agent itself does not build the gall.

Gall induced by a scale insect, *Cystococcus pomiformis*, on a *Corymbia* sp. Photo: Emma Woodward.

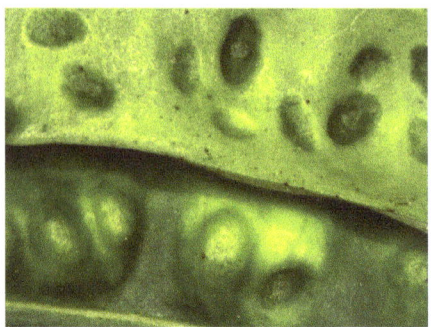

Leaf pit galls of a psyllid, *Trioza eugeiae*, on *Syzygium paniculatum*. The top half of the picture shows the upper side of a leaf and the lower half shows the underside of a leaf. The insects have left the galls. Photo: Gary Taylor.

Galls induced by the nematode worm *Anguina microlaenae* on a species of grass. Photo: Ian Riley.

The study of plant galls is called cecidology. This name is derived from the Greek word *kekis,* meaning a gall.

Most plant galls are the result of insect activity. However, some protozoa, rotifers, viruses, fungi, bacteria, nematode worms and mites, as well as physical injury, can also disrupt normal plant growth and cause galls to form. Insect-induced galls can be as simple as the leaf pits induced by some plant-sucking bugs, or as complex as the enclosed woody stem galls induced by some scale insects. Most galls are inhabited by immature insects (larvae or nymphs) but some, for example galls induced by thrips, also house adults.

Unlike galls induced by other organisms, the form and structure of an insect-induced gall is usually characteristic of the insect species that caused it. Sometimes the galls of males and females of the same insect species look different (sexual dimorphism) and some insect species induce different looking galls at different times in their life cycle.

Gall of a psyllid, *Schedotrioza multitudinea*, on *Eucalyptus obliqua*, cut open to show the nymph. Photo: Gary Taylor.

A stem gall on *Eucalyptus globulus* cut open to reveal the larvae of the chalcidoid wasp, *Selitrichodes globulus*, that induced the gall. Photo: Gevork Arakelian.

A stem gall caused by the rust fungus *Uromycladium tepperanium* on an *Acacia* sp. Photo: Eric Frei.

Stem gall of a thrips, *Iotatubothrips* sp., on a *Casuarina* sp., cut open to show the adults (with dark stripes) and the pale larvae. Photo: Laurence Mound.

Two female galls and one male gall of the scale insect *Apiomorpha minor* on a eucalypt. The male gall is the small brown tubular leaf gall in the centre. The female galls are the much larger green stem galls.

Unicelled leaf galls induced by the psyllid *Schedotrioza multitudinea* on *Eucalyptus obliqua*. The adult insects have departed and only their old nymphal skins remain near the exit holes in the galls. Photo: Gary Taylor.

Galls induced by the wasp *Nambouria xanthops* on a eucalypt leaf. The two pink ridges are female galls and the male gall is the slightly expanded leaf vein near the lower edge of the leaf between the female galls. Photo: Jonathan Barran.

Galls can have a single chamber (unicelled) or numerous chambers (multicelled). Each chamber may contain a single individual insect or many individuals. The structure of gall chambers can vary but most include an inner zone rich in nutrients, such as nitrogenous compounds and soluble sugars, surrounded by a stronger supporting wall enclosed by the plant's outer layer of cells (epidermis). There may also be a layer where plant components that are difficult for insects to digest are secreted away from where the insect is feeding.

A multicelled stem gall induced by the chalcidoid wasp *Selitrichodes globulus* on *Eucalyptus globulus*. The gall has been cut open to show its numerous chambers. Photo: Gevork Arakelian.

How important is the right host plant?

Each species of gall-inducing insect is generally restricted to living on one, or a few closely related host species. Consequently, when a host plant evolves to form new plant species, an insect that induces galls on the original host plant may not be able to colonise the new plant hosts or may not survive and reproduce as well on the new plant hosts as they do on the original plant host. Over time, evolutionary pressures acting on the insects may result in the development of new gall-inducing insect species adapted to the new host plant species.

Insect-induced galls can form on any part of a plant but many gall-inducing insect species are able to induce galls only on one part of the plant, commonly on the leaves. Actively growing plant tissue, such as that found in young leaves, stems and flower buds, is usually required for a gall to be induced. Only rarely are galls initiated on mature plant parts.

Galling occurs in most plant groups worldwide, but appears to be most common in vegetation with sclerophyllous (hard) leaves or in places that experience wet-dry seasonal environments. Fossil evidence of insect galls on tree fern fronds suggests that insects have been inducing galls on plants since at least 300 million years ago, during the Late Pennsylvanian times of the Late Carboniferous.

The number of times in a year that a gall-inducing insect species produces offspring varies from one to many, depending on the insect species, the host plant species, the plant part involved and the prevailing environmental conditions. This variability is well illustrated by gall-inducing midges.

Midge species that induce galls on trees usually only have one generation per year (univoltine) because most trees produce new shoots and flowers once a year. Midges on herbaceous plants that produce new shoots repeatedly can have many generations per year (multivoltine). A few midge species achieve two generations a year (bivoltine) by alternately using different organs on the same host plant species.

The effect of environment on the number of generations per year is evident in a widespread psyllid species from India. This species produces six to eight generations per year when it occurs in a subtropical area but only one generation per year when it occurs in a temperate area. In both areas the host plant species is the same.

Stem galls induced on Molloy red box, *Eucalyptus leptophleba*, by an unidentified wasp species.
Photo: John Ludwig.

The sclerophyllous leaves of a eucalypt, *Eucalyptus cinerea*. Photo: Il-Kwon Kim.

How do insects cause a plant to form a gall?

One of the most tantalising questions relating to galls is: 'How do the insects do it?' The short answer is: 'No-one knows for sure'.

We do know that in some cases a gall begins to form when an insect lays an egg in plant tissue. In many other cases it is the feeding activity of the insect, usually an immature stage, which initiates gall formation. We have also learned that the continuing presence of the insect, especially its feeding activity, is required for the gall to remain alive and grow. From this we can assume that chemicals injected with the egg and/or the saliva of a gall-inducing insect are responsible for gall development, and that chemicals in the insect's saliva (and sometimes in its frass and glandular secretions) maintain the gall.

What we don't know is what these chemicals are or how they work. Chemicals that have been implicated include phenolic compounds, free amino acids and plant growth promoting chemicals such as auxins and cytokinins. It is even possible that initiation of a gall results from the transfer of genetic material (DNA or RNA) from the insect to the plant. This genetic material could then redirect the plant's cells to start making a gall. Gall-inducing insects may be natural genetic engineers!

How does living in a gall benefit an insect?

There are several possible benefits to be gained by an insect living in a gall. The most obvious advantage is better access to a nutritious food supply, free from hard-to-digest plant chemicals such as oils and tannins.

The gall may also provide protection from animals, birds and insects that would otherwise prey on the gall-inducing insect. However, protection is not guaranteed because some insects (known as parasitoids) specialise in developing in, or on, the body of the gall-inducing insect while it is in the gall. Other insects (inquilines) invade the gall chamber and force the gall-inducing insect to share its home with these intruders. There are also thieving insects (kleptoparasites) that take over the gall and displace the original gall-inducing insect. In addition, some animals are not deterred by tough walls and will chew holes in them to get at the insects or the nutritious cells inside the gall.

Another potential advantage to an insect of living in a gall is protection from drying out in arid environments. The evidence for this benefit is inconclusive at present and does not appear to apply in Australia.

How do galls affect their host plants?

Because plant nutrients are concentrated in and around galls they are said to act as 'nutrient sinks'. This is a disadvantage for the plant because it means that nutrients, which would otherwise be used by the plant to grow new plant parts for itself, become food for developing insects instead. Galls that form on flower buds or seeds can reduce the amount of fruit and seed produced by the plant. On the other hand, some gall-inducing insects, for example fig wasps, play an essential role in the transfer of their host plant's pollen to fertilise its flowers and so are vital to the survival of that plant species.

The importance of insect-induced galls

The importance of galls may not be immediately obvious to a casual observer because many galls, and the insects that cause them, are tiny and appear insignificant. However, in reality, their importance far outstrips their size and abundance. Gall-inducing insects that pollinate plants, or reduce seed production by galling flower buds or seeds, can have a strong influence on the composition of plant communities. In addition, galls often house many other organisms as well as the gall initiator. These non-galling organisms may depend wholly, or partly, on the gall for survival. Thus gall-inducing insects play a critical role in maintaining the structure of their ecological community.

If you had lived in the Middle Ages you would probably have regarded insect-induced galls as supernatural growths and consulted them to predict the future. If a gall contained a grub-like larva you would have expected the next year to bring famine; if it housed a 'fly' then war was imminent; and if there was a spider inside, then a devastating disease epidemic was coming. Such beliefs were held in the Western world for several centuries and continued even after the 17th century when the Italian doctor Marcello Malpighi established the true origin of insect-induced galls.

Galls have been valued for their more practical qualities too. In the past people have used gall extracts to tan hides, clean teeth, provide bait for fishing, make printing ink and dyes, treat skin and eye infections, and make jewellery. Galls and their insects have been, and sometimes still are, a source of food for both livestock and humans.

Gall-inducing insects can be pests and cause economic losses in horticulture, agriculture and forestry but because of their tight associations with only one, or a few closely related, host plant species, they also have the potential to be effective control agents for invasive plant species.

Today we know that insect-induced galls can be ideal experimental subjects in scientific research. Their usefulness for this purpose stems from the enclosed nature of their lifestyle and the close associations they have with their host plants. Scientists now employ gall-inducing insects to search for answers to questions in a diverse range of topics including: plant physiology, insect nutrition and reproduction, complex community interactions, genetics, behaviour, ecology and evolution.

Gall-inducing insects and their host plants

The biology and ecology of most Australian gall-inducing insect species are poorly understood and taxonomic studies are incomplete. Consequently the number of insect and plant species associated with Australian galls is likely to increase in the future as our knowledge grows.

Gall-inducing bugs

Some of the best studied and most conspicuous gall-inducers in Australia are scale insects of the superfamily Coccoidea (Hemiptera: Sternorrhyncha). There are about 180 known Australian species of these bugs that induce galls in stems, leaves and flower buds. Many of their galls look like enclosed swellings, or tubes, with a single external opening (orifice) but others resemble blisters, pits, pouches, rosettes or buds. Most occur on eucalypts (*Eucalyptus* and *Corymbia* species) or other members of the plant family Myrtaceae but coccoid galls

are also known on banksias (Proteaceae), sheoaks (*Allocasuarina* and *Casuarina* species) (Casuarinaceae), acacias (Mimosaceae), hoop and kairi pine (Araucariaceae), and beeches (Fagaceae). Scale insects of the genus *Apiomorpha* are responsible for causing Australia's largest galls. Although the tubular galls of males of this genus are no more than 1 cm long, the

This gall of a scale insect, *Apiomorpha strombylosa*, has been cut open to show the adult female. The rear end of the insect is pointing to the left. The insect is on its back so you can see its tiny legs.
Photo: Lyn Cook.

Female galls induced by a scale insect, *Apiomorpha malleeacola,* on eucalypt gumnuts.
Photo: Penny Gullan.

A young two-horned female gall of the scale insect *Apiomorpha munita* on a eucalypt stem.
Photo: Penelope Mills.

Two female galls induced by a scale insect, *Frenchia casuarinae*, on the stem of a sheoak, *Allocasuarina verticillata*. Photo: Penny Gullan.

An old dry specimen of the female gall of the scale insect *Apiomorpha duplex* on a eucalypt stem. This is one of the largest Australian galls. Photo: Penelope Mills.

Large rosette-like female galls and thin stem-like male galls of the scale insect *Cylindrococcus spiniferus* on the stem of a sheoak, *Allocasuarina littoralis*. The female galls are often mistaken for the fruit of the tree. Photo: Penny Gullan.

chamber of the galls of females can be up to 11 cm long and can have additional spectacular blade-like appendages that are 14 cm long.

There are also numerous Australian gall-inducing psyllid bugs in the superfamily Psylloidea (Hemiptera: Sternorrhyncha). Most galls are on leaves and can be in the form of pits, leaf rolls or raised globular sacs. Most reported host plants are eucalypts (*Eucalyptus* and *Corymbia* species) (Myrtaceae) but galls of these bugs can also be found on lilly pillies (*Syzygium* and *Acmena* species) (Myrtaceae), figs (Moraceae), banksias (Proteaceae), *Geijera* species (Rutaceae), beach mahogany, *Calophyllum inophyllum* (Guttiferae), brush box, *Lophostemon confertus* (Myrtaceae), pale turpentine bush, *Beyeria lechenaultii* (Euphorbiaceae) and the kamala tree, *Mallotus philippensis* (Euphorbiaceae).

Some Australian whiteflies of the superfamily Aleyrodoidea (Hemiptera: Sternorrhyncha) are gall-inducers. These tiny bugs induce pits which are usually 2–3 mm in diameter and coloured red, pink or yellow. One species is reported to cause pit galls on the leaves of dryandras and banksias (Proteaceae). Another is found on acacias (Mimosaceae) that have no true leaves but instead have leafstalks (petioles) modified to function like leaves

(phyllodes). This whitefly species produces deep pits on the surface of the phyllodes.

Leaf roll gall of a *Trioza* sp. psyllid on pale turpentine bush, *Beyeria lechenaultii*. Photo: Richard Glatz.

An adult female psyllid, *Schedotrioza multitudinea*, on *Eucalyptus obliqua*. Photo: Gary Taylor.

The eggs of the psyllid *Schedotrioza multitudinea* on a leaflet of *Eucalyptus obliqua*. Photo: Gary Taylor.

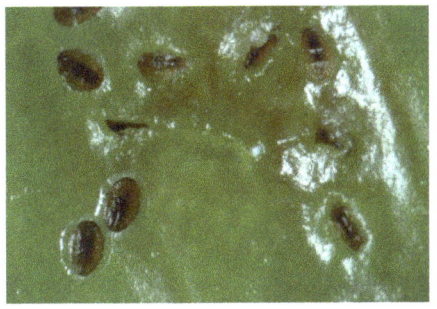

First instar nymphs of the psyllid *Schedotrioza distorta* initiating galls on upper surface of a leaf of *Eucalyptus leucoxylon*. Photo: Gary Taylor.

Pupae of the whitefly *Viennotaleyrodes lacunae* on the underside of a phyllode of *Acacia rubida* (dried museum specimen). Photo: Jon Martin.

Gall-inducing wasps

In Australia, the wasp superfamily Chalcidoidea (Hymenoptera: Apocrita) contains many gall-inducing species. The chalcidoid group includes the families Agaonidae, Eulophidae, Eurytomidae, Pteromalidae, Tanaostigmatidae and Torymidae. Many of these tiny wasps induce galls in the stems, leaves or seeds of eucalypts (*Eucalyptus* and *Corymbia* species) (Myrtaceae) but others utilise the developing florets of figs (Moraceae) or the flower buds of acacias (Mimosaceae) or hakeas (Proteaceae). Two chalcidoid species induce galls in the fruits of the desert lime, *Eremocitrus glauca* (Rutaceae), and one deforms the stems of Geraldton Wax, *Chamelaucium uncinatum* (Myrtaceae).

A relatively recent discovery is that wasp species of the family Braconidae, belonging to the superfamily Ichneumonoidea (Hymenoptera: Apocrita), can also be gall inducers. Four species of gall-inducing braconid wasp are known in Australia. They cause leaf or stem galls on banksias (Proteaceae).

Gall-inducing thrips

More than 27 species of native thrips, family Phlaeothripidae (Thysanoptera: Tubulifera) are known to induce galls. Their galls are usually on leaves and are described as leaf rolls, tubes or pouches. Some are covered with spikes, hairs or ridges and some are divided internally into numerous interconnecting compartments. They are often associated with the group of acacias (Mimosaceae) that have phyllodes instead of true leaves. Plant genera that are known to have thrips galls on true leaves include *Pittosporum* (Pittosporaceae), *Auranticarpa* (Pittosporaceae), *Myoporum* (Myoporaceae), *Olearia* (Asteraceae),

An adult female chalcidoid wasp, *Selitrichodes globulus*, that induces stem galls on *Eucalyptus globulus*. Photo: Gevork Arakelian.

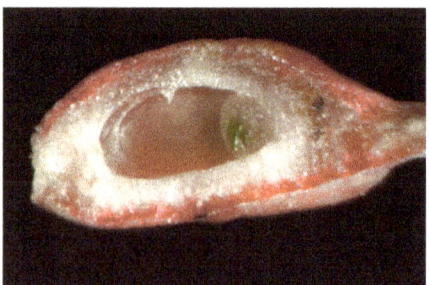

This leaf gall of the chalcidoid wasp *Ophelimus maskelli* has been cut open to show the larva inside. Photo: Zvi Mendel and Alex Protasov.

Galls induced by a wasp, *Ophelimus* sp., on *Eucalyptus cinerea*. Photo: Il-Kwon Kim.

This gall of the thrips *Kladothrips rodwayi* on a phyllode of *Acacia melanoxylon* has been cut open to show an adult and eggs on the inner surface of the gall. Photo: Laurence Mound.

A thrips gall cut open to show the larvae. Photo: Laurence Mound.

A pouch gall of the thrips *Kladothrips rugosus* on a phyllode of *Acacia pendula*. Photo: Laurence Mound.

Thrips galls on the leaves of an *Auranticarpa* sp. Photo: Laurence Mound.

The leaves of *Myoporum insulare* on the left are normal, while those on the right have galls induced by a thrips. Photo: Laurence Mound.

Thrips-induced galls on the leaves of *Pittosporum undulatum*. Photo: Laurence Mound.

Callistemon (Myrtaceae), *Geijera* (Rutaceae) and *Ficus* (Moraceae). Some thrips induce woody galls inside the slender branchlets of sheoaks (Casuarinaceae). There are no thrips galls recorded from eucalypts.

Gall-inducing flies

Australian flies known to induce galls belong to four families: Cecidomyiidae (Diptera: Nematocera) and Tephritidae, Agromyzidae and Fergusoninidae (Diptera: Brachycera). There is currently no reliable estimate of the number of gall-inducing fly species but it is likely to be high. Fly galls occur mainly on stems, leaves and flower buds. Their hosts include many plant species from numerous families. For example, cecidomyiid midges induce galls on banksias (Proteaceae), cypress-pines (Cupressaceae), black palms (Arecaceae), sheoaks (Casuarinaceae), eucalypts (*Eucalyptus* and *Corymbia* species) (Myrtaceae), tea tree, (*Leptospermum* species) (Myrtaceae), acacias (Mimosaceae), hop bush (Sapindaceae), glassworts (Chenopodiaceae), grey mangrove,

A female fergusoninid fly, *Fergusonina* sp., laying its eggs on a bud of a *Melaleuca* sp. Photo: Susan Wineriter, USDA Agricultural Research Service, Bugwood.org.

The adult midge *Asphondylia floriformis* induces galls on beaded glasswort, *Sarcocornia quinqueflora*. Photo: Anneke Veenstra-Quah.

Galls induced by a fergusoninid fly and its associated nematode worm species on flower buds of a *Melaleuca* sp. Photo: Susan Wineriter, USDA Agricultural Research Service, Bugwood.org.

A cross-section of a gall induced on shrubby glasswort, *Tecticornia arbuscula*, by an *Asphondylia* midge reveals the gall chambers and a midge pupa. Photo: Anneke Veenstra-Quah.

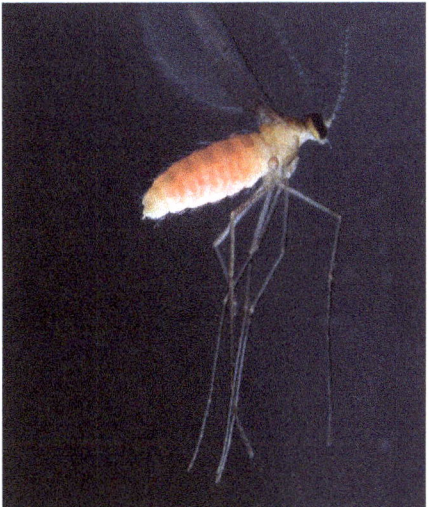

The midge *Lophodiplosis trifida* induces galls on the paperbark *Melaleuca quinquenervia*. Photo: Susan A Wright USDA ARS Invasive Plant Research Laboratory.

A pupa of the gall-inducing midge *Lophodiplosis trifida*. Photo: Susan A Wright USDA ARS Invasive Plant Research Laboratory.

This gall on shrubby glasswort, *Tecticornia arbuscula*, has been induced by an *Asphondylia* midge. Photo: Anneke Veenstra-Quah.

Avicennia marina (Acanthaceae), and *Solanum* species (Solanaceae). Fergusoninid flies are found on eucalypts (*Eucalyptus*, *Corymbia* and *Angophora* species) (Myrtaceae) as well as on species of paperbark, (*Melaleuca* species) (Myrtaceae), while tephritid fly galls are known from plants in the families Asteraceae, Heliantheae and Goodeniaceae.

Gall-inducing beetles

Few Australian beetles are gall inducers. We know of six species in the family Buprestidae (Coleoptera: Polyphaga). These jewel beetle species each cause galls on stems or roots of either sheoaks (*Allocasuarina* species) (Casuarinaceae), *Pultenaea* species (Fabaceae) or *Dillwynia* species (Fabaceae). Beetle species in the subfamily Sagrinae of the family Chrysomelidae (Coleoptera: Polyphaga) are thought to live in stem galls of various plants, including native yams (Dioscoreaceae).

A stem gall induced by the jewel beetle *Ethonion leai* on an orange parrot-pea. The gall on the right has been cut open to reveal the larva. Photos: Peter Lang.

The gall-inducing jewel beetle *Ethonion leai* on the flowers of its host plant *Dillwynia hispida*, the orange parrot-pea. The beetle is 6.5 mm long. Photo: Peter Lang.

Gall-inducing moths

Gall-inducing moths are rare in Australia. One species of the genus *Coleophora* in the family Coleophoridae (Lepidoptera: Glossata) is reported to form stem galls on chenopod plants, family Chenopodiaceae and one species of *Alucita*, family Alucitidae (Lepidoptera: Glossata) causes large elliptical galls on the stems of *Canthium* species (Rubiaceae). *'Tinea' vetula*, family Incurvariidae (Lepidoptera: Glossata) forms pear-shaped galls on the leaves of *Banksia integrifolia* (Proteaceae).

An adult of a gall-inducing moth, *Alucita* sp., resting on its gall. Photo: Murray Upton.

Here the gall of *Alucita* has been cut open to show the larva and its frass. Photo: Murray Upton.

Gall inducing insects around the world

Galls induced by bugs, wasps, thrips and flies are common throughout the world. However, there is generally a paucity of gall-inducing beetles and moths, including in Australia. The main difference between Australia and some other parts of the world is the absence of native gall-inducing aphids, superfamily Aphidoidea (Hemiptera: Sternorrhyncha) and sawflies, superfamily Tenthredinoidea (Hymenoptera: Symphyta). There is also an almost complete absence of native gall-inducing cynipid wasps from the super-family Cynipoidea (Hymenoptera: Apocrita). These three groups contain many gall-inducing species in North America and in northern and central Europe.

Galls induced on leaf petioles of a poplar, *Populus* sp., by the aphid *Pemphigus populitransversus*. Photo: Herbert Pase, Texas Forest Service, Bugwood.org.

Galls on the underside of a leaf of blue oak, *Quercus douglasii*, induced by the cynipid wasp *Andricus brunneus*. Photo: Ron Russo.

Galls induced by the sawfly *Pontania virilis* on purple willow, *Salix purpurea*. Photo: Tommi Nyman.

An adult female sawfly, *Euura lanatae*, that induces galls on woolly willow, *Salix lanata*. Photo: Tommi Nyman.

The oriental chestnut gall wasp, *Dryocosmus kuriphilus*, is a cynipid wasp.
Photo: Gyorgy Csoka, Hungary Forest Research Institue, Bogmood.org.

Susceptible plants and their distribution

Fewer than 50 of the more than 2250 Australian vascular plant genera are reported to support native gall-inducing insect species. More than 50 per cent of these insect species occur on eucalypts (*Eucalyptus*, *Corymbia* and *Angophora* species) (Myrtaceae). Approximately 18 per cent occur on either acacias (Mimosaceae) or figs (Moraceae). The rest are unevenly distributed among the remaining gall-susceptible plant genera, none of which has more than three per cent of the total gall-inducing insect fauna. Consequently, most Australian plant species have no known insect-induced galls.

The concentration of galling in only a few plant groups occurs in other countries too but the galled plant groups are different. For example, in South America, Africa and India leguminous plants (Fabaceae) are most commonly attacked, whereas in Europe and North America approximately half the gall-inducing insect species are found on oaks (Fagaceae).

The main constraint on the geographic distribution of each gall-inducing insect species is the geographic distribution of its host plant species. Because the majority of insect-induced gall species occur on eucalypts, acacias and figs, which are all very widespread groups, the incidence of galling in Australia is also widespread.

Eucalypts can be found in almost all environments but they are the dominant canopy species on infertile soils of temperate and subtropical coastal regions.

More than 50 per cent of Australian gall-inducing insect species occur on eucalypts. Photo: John Ludwig.

Acacias also occur in numerous habitats but are particularly prevalent in arid, semi-arid and dry sub-tropical regions. Figs are rainforest species. Rainforests occur in patches from the Kimberley region in northern Western Australia, across the Northern Territory to Cape York in Queensland, and down the east coast of Australia to Tasmania.

Most of the 700 or more species of eucalypt are native only to Australia. There are almost 1000 native acacia species and more than 40 Australian fig species. The large numbers of species in these far-ranging gall-susceptible plant groups provide Australian gall-inducing insects with a vast array of potential hosts and environments.

Remarkable adaptations

An insect that spends part of its life in a gall has the advantages of a relatively safe home and nutritious food. However, it also has to cope with some unique challenges not faced by insects that are free-living for their entire lives. For instance, how do gall insects get inside the appropriate plant tissue in the first place and how do they get out again? How do they gain access to their food or find a mate and reproduce?

Gall-inducing insects have evolved in some remarkable ways in order to deal with the difficulties associated with an enclosed lifestyle.

Finding the right site

Finding the right host plant and choosing the right part of the plant on which to initiate a gall is vitally important for a gall-inducing insect because its relationship with its host plant is so close. Usually each insect species can only influence one, or a few, plant species to produce galls and then usually only on certain plant parts and at a time when the plant cells are rapidly dividing. Some scale insect species are able to initiate galls on the surface of other mature scale galls, and there is an armoured scale that appears to initiate pit galls on almost mature leaves, but these are unusual among gall-inducing insects.

Some gall-inducing insects are known to have specialised sensory organs or appendages on their antennae, mouthparts, legs or abdomen for detecting chemical cues that lead them to their gall sites. For example, female chalcidoid wasps that gall flowers of figs rely on volatile chemicals,

A female fig wasp, *Pleistodontes* sp. Photo: Paul Zborowski.

specific to each fig, that signal when the developing flowers are at a suitable stage. The modified antennae of these wasps have numerous sensory processes on segments toward the tips of the antennae that are capable of detecting the chemicals released by the host fig. Other factors that are likely to be taken into account by searching insects are the nutritional status of the plant tissues, the presence of secondary plant chemicals (metabolites such as oils and tannins) and the surface structure, colour, shape and size of the plant parts.

Life cycles of gall-inducing insects

Most Australian gall-inducing insects have a life cycle that includes the stages of egg, larva, pupa and adult. After the larva hatches from the egg it grows and passes through several developmental stages (instars) before becoming a pupa. Usually pupae do not feed or move, although some fly pupae can jump. During the pupal stage the insect undergoes a dramatic dissolution and rearrangement of its cells (complete metamorphosis) before finally adopting the form and functions of an adult. Thrips do not actually undergo complete metamorphosis but pass through a similar stage in which they do not feed and are mostly immobile.

The life cycles of three gall-inducing groups: psyllids, scale insects and whiteflies, are simpler. They include only an egg, nymph and adult stage. Once out of the egg, the nymph grows and sheds its cuticle layer several times as it gradually acquires adult characteristics.

In flies, psyllids, whiteflies, wasps, beetles and moths, it is the adult female that finds the appropriate plant species and, either by depositing eggs onto the plant surface, or within the plant tissues, ensures that her offspring will be near, or in, the right place to induce a gall. When the young insect hatches from the egg its feeding activity and salivary secretions trigger gall formation, which causes the gall to grow.

It appears that chemicals introduced during egg-laying, or the egg itself, can sometimes also play a part in gall initiation. This has not been established conclusively for Australian gall-inducing insects but seems likely for whiteflies because their eggs have been observed enclosed in leaf pits before the nymphs have emerged from the egg.

Adult thrips also choose the gall site but in this case it is the feeding activity of either a single adult female thrips, or a male and female together, that initiates gall formation. Sometimes adults engage in fights with other thrips of the same species over possession of prospective gall sites. The combatants attack one another, often fatally, with their enlarged, armed forelegs. The rate of gall development in some thrips species

A gall-inducing chalcidoid wasp, *Ophelimus* sp., deposits its eggs on a stem of *Eucalyptus cinerea*. Its needle-like ovipositor extends from its abdomen into the stem. Photo: Il-Kwon Kim.

A woody gall induced by the thrips *Phallothrips houstoni* on *Casuarina cristata*. The large orange and brown striped individuals are wingless adults and the smaller white ones are larvae. There are also a few smooth-surfaced creamy eggs visible. Photo: Laurence Mound.

apparently decreases or increases as the plant growth rate slows down or accelerates in response to variations in temperature and rainfall.

In scale insects it is usually the youngest immature insect stage (first instar nymph or crawler) that disperses from the maternal gall to find a new gall site. The galls of males can resemble the galls of females of the same species or they can look very different. Some Australian scale insect species have developed an unusual way of facilitating the dispersal process. In these species the young males mature faster than the females (sexual dichronism) and the males reach adulthood while still inside the maternal gall. Once they are mature the winged males transport their younger sisters out of the maternal gall to sites that are suitable for the females to feed and initiate new galls. The immature

An adult male scale insect, *Cystococcus pomiformis*, with his first instar sisters clinging to his abdomen. Photo: Penny Gullan.

females have no wings (apterous) and cling to the elongated abdomens of their older brothers while being transported. This association is termed phoresy.

In Australia, adult gall-inducing beetles, moths, flies and wasps are all free living but their grub-like larvae develop within a gall and then either pupate in the gall or emerge and drop to the ground to pupate in the soil.

Some beetles that pupate within a gall do so in specially constructed cocoons. Galls of moths usually have a single immature insect in each gall but galls of beetles, flies and wasps can contain a single individual or numerous individuals, each in its own chamber within the gall.

A thrips gall may contain numerous individuals and include adults and immature insects. There are two larval stages and two or three 'pupal' stages.

This gall of the thrips *Kladothrips rugosus* has been cut open to show a physogastric female thrips (top centre). Note the jelly-like appearance of the physogastric female's abdomen. There are also small white eggs and larger white larvae scattered around the inside of the gall. The adult individual lower in the picture is a male thrips. Photo: Laurence Mound.

Pupation occurs either within the gall or in the soil. In some species the gall-inducing female develops a greatly enlarged abdomen (becomes physogastric) and produces a single generation of up to 1000 larvae.

A single developing psyllid nymph inhabits each psyllid gall. The nymph passes through five developmental stages (instars) within the gall before it is mature. The adults are free living and resemble minute cicadas. Gall-dwelling psyllid nymphs usually have shorter legs and antennae and smaller eyes than free-living psyllid nymphs and many are partly or entirely covered in waxy secretions. Psyllid nymphs that live in simple pit galls tend to be oval or circular in shape and flattened dorsally. While the ventral surface of the insect is relatively soft, the dorsal surface is shield-like, hard and dark (sclerotised). It acts as a seal to plug the pit opening. The bodies of psyllids living in more complex, closed galls are only weakly sclerotised and are 'inflated', not flattened.

Little is known about the biology of gall-inducing whiteflies except that a single nymph develops in each pit gall. The adults are free living.

This gall induced by the psyllid *Schedotrioza cornuta* on *Eucalyptus socialis* has been cut open to show the developing nymph. Its body is flattened dorsally and has a waxy covering. Photo: Gary Taylor.

An adult female psyllid, *Schedotrioza serrata*. Photo: Gary Taylor.

In those scale insects species in which male nymphs are known to form galls each gall is inhabited by a single developing male until it becomes an adult. Adult males have one pair of simple wings and are fragile and short-lived but adult females are larger and more robust.

The galls of female scale insects house the female while she completes development and, eventually, the mature female and her developing offspring. Some adult females can live in their galls for more than a year. Female scale insects that live in the most enclosed galls often plug the hole at the exit to the gall until they are mature. The plug can be formed from a modified and usually sclerotised part of the abdomen, wax secretions or shed skins from the female's body, or decaying plant tissue.

Male scale insect nymphs have four developmental stages but female nymphs have only two or three. Adult female scale insects are sometimes described as neotenic because they have passed through fewer developmental stages than males and are able to reproduce while still appearing immature. They usually have no wings, reduced eyes, antennae and mouthparts, and fewer body secretions than free-living scale species. All but a few species also have reduced legs.

Finding food

While they are developing in the gall, most immature larvae or nymphs feed on the highly nutritious inner lining of cells that usually forms inside galls. The way they feed varies with the kind of mouthparts they have. Wasp, beetle, moth and some fly larvae have mouthparts that allow them to chew the plant cells. The frass produced by beetle and moth larvae feeding is often stored within the gall in packed form.

This gall induced by the jewel beetle *Ethonion leai* has been cut open to show the waste products (frass) stored in the lower part of the gall.
Photo: Peter Lang.

This gall induced by the scale insect *Apiomorpha strombylosa* has been cut open to show the neotenic adult female. The insect is on its back with its head pointing toward the bottom left of the picture.
Photo: Penny Gullan.

Some fly larvae have parallel rows of retractable mouth hooks capable of sweeping and digging, which they use to 'mow' plant cells before swallowing them. The larvae of fergusoninid flies usually have a specialised, sometimes heavily sclerotised, structure on their backs called a dorsal shield. It is thought this structure may be used by some species to scrape cells from the internal surface of the gall to stimulate the formation of new nutritive tissue.

Some midges have a larval feeding strategy that is very different from other gall-inducers. The galls of these midges do not develop the nutritive layer found in galls of other insect species so, after wounding plant cells to initiate gall formation, the larvae feed for a while on plant tissue. They then switch to feeding mainly on a fungus that grows on the inner walls of the gall. When newly emerging adult females leave the gall they carry fungal spores with them to establish the food

supply for the next generation. The spores are thought to be carried either in the gut or on a modified body segment on the underside of the abdomen. They are introduced into the gall site when the female is laying eggs into plant tissue with her needle-like ovipositor.

Psyllid and scale insect nymphs use their stylets – threadlike double-channelled mouthparts – to probe plant cells for their food. Thrips larvae and adults feed by sucking the contents of individual plant cells with their asymmetric, single-channelled mouthparts.

Adult male scale insects have no functional mouthparts and so do not feed but most adult females tap into plant tissues that transport organic food materials from the leaves. Honeydew, the sugary waste produced by phloem feeding, is ejected out through the exit hole of the gall where it is regularly collected by ants. If not removed honeydew could promote the growth of

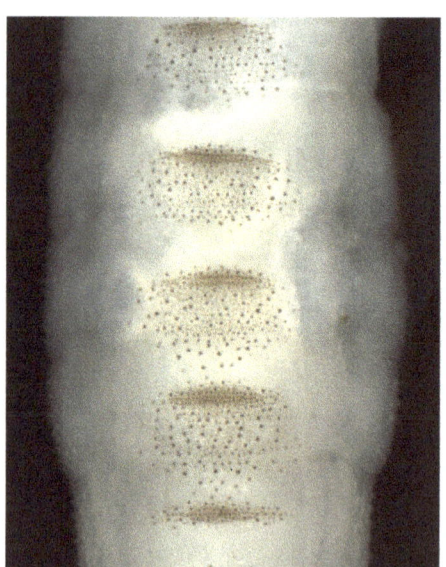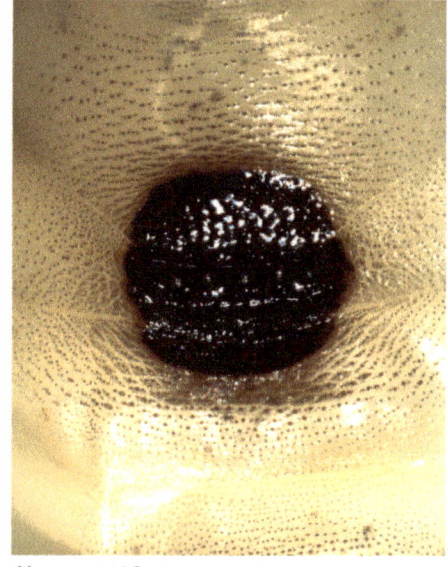

The dorsal shields of the larvae of two different species of fergusoninid fly. Photo: Gary Taylor.

A gall induced by the midge *Asphondylia floriformis* on beaded glasswort, *Sarcocornia quinqueflora*. The gall has been cut open to show the larva and the white fungus lining the inner walls of the chamber. Photo: Teresa Lebel.

micro-organisms inside the gall and harm the inhabitants.

There are few reports of the feeding habits of other adult gall-inducing insects. Some adult beetles eat flower petals and stamens and some adult wasps are known to feed on nectar from flowers. Adult midges may drink water from dew (although they don't eat food).

The adults of some psyllid species have an unusual feeding behaviour. They appear to prefer to feed on the convex surfaces of the pit galls induced by their nymphs.

Reproduction

Most adult gall-inducing insects live outside the gall and mating takes place shortly after they become adult. For example, adult male braconid wasps have reduced wings that limit their flying abilities. They stay close by after exiting their galls and are believed to

mate on the galls as the adult females emerge.

In other countries male gall midges that are similar to Australian species swarm near females on the host plant before mating. The males are attracted by the sex pheromones released by the females. The female midges mate only once but some males mate several times. Australian midges are likely to have similar mating behaviours.

Mating in psyllids is preceded by the male dancing around the feeding female and producing sounds that appear to be made by vibrating structures on its wings and thorax. When she is receptive the female indicates this by flexing her wings and the end of her abdomen. She may also release a sex pheromone.

The only Australian gall-inducing insects that mate before the female is out of the gall are fig wasps, scale insects and some thrips. Both male and female fig wasps develop in separate galls within the same fig. As soon as a male fig wasp escapes from its gall into the fig cavity it looks for a gall containing a female, gnaws a hole in the gall and uses its telescopic genitalia to mate with the female inside.

In most scale insects the free-living adult male must use its elongated abdomen, or its long, thin, sclerotised or elastic genitalia, to reach into the mature gall and mate with the female inside. Some female scale insects have long hind legs that may be used to grasp and guide the male abdomen during mating.

In thrips that pupate in the gall the newly emerged males and females both inhabit the same gall. They mate within two or three days of becoming adults.

Leaving the gall

The way an insect leaves its gall depends on the thickness of the gall wall and how enclosed the gall is. The method also varies depending upon the kind of insect and whether it is an adult or immature when it leaves the gall. For example, thrips galls are usually incompletely sealed and often open further at maturity. Dispersing adult thrips (or larvae in species that pupate in the soil)

exit through the natural openings of the gall. Psyllid nymphs leave their galls at the final nymphal stage either by unblocking a naturally occurring pore in the gall wall or via the opening formed when the gall itself splits open as the infested leaf grows. Psyllid nymphs moult into the adult stage on the surface of leaves. Conversely, the galls of moths are completely enclosed and the larva, before pupating in the gall, usually prepares the exit opening for the adult by weakening part of the gall wall.

In some midges the mature larva forces its way out of the gall, drops to the ground and jumps around in search of a suitable place in the soil for it to pupate. In midges that pupate in the gall either the fully mature larva uses a specialised, elongate epidermal structure to cut a tunnel almost through to the outside (to be completed later by the mature pupa) or the mature

The pupal case of the gall-inducing midge *Asphondylia floriformis* after the adult has emerged. The pupa was supported by the gall while the adult was developing. Photo: Anneke Veenstra-Quah.

This mature psyllid gall has split open. It was induced by *Schedotrioza cornuta* on a leaf of *Eucalyptus socialis*. Photo: Gary Taylor.

pupa cuts the entire exit tunnel. In both cases the mature pupa only makes a partial exit and is supported by the gall while it transforms into an adult. Fergusoninid flies pupate within the gall and the adult uses an expandable pouch (ptilinum) on its head to break through a thin area of the gall wall and escape.

Wasps pupate in their galls and the adults chew their way out. Fig wasps must also escape from the fig. Many adult male fig wasps do not survive for long outside the fig because they are specialised to walk around on the uneven surface inside the fig and are unstable on flat surfaces. In spite of this, they start digging their way out of the fig after mating is finished. This behaviour is important for the survival of the next generation of these wasps because it creates a tunnel for the winged females to exit the fig and disperse to find a new fig to lay eggs in.

Most adult female scale insects never leave their galls but their offspring exit via the natural opening at the end of the gall. Adult male scale insects that form tubular galls escape by simply backing out of their open-ended galls.

Complex relationships

Most Australian gall-inducing insects appear to be self-sufficient but a few require the presence of other organisms, apart from their host plants, in order to survive. Fungus-feeding gall midges are one example of such a relationship. The midges need the fungus for food and possibly for assistance in initiating the gall. The fungus may or may not benefit from the relationship.

When both organisms in a relationship benefit from the association it is described as a mutualistic relationship. Fergusoninid

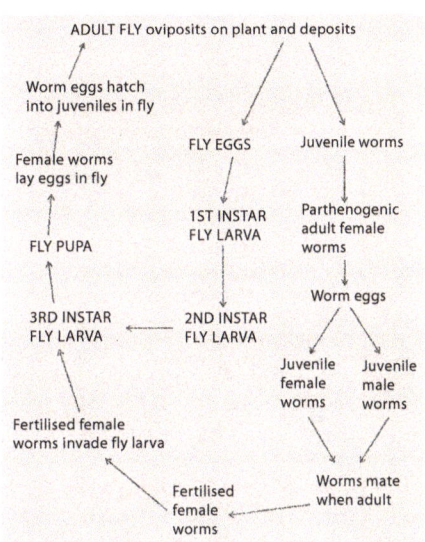

The life cycle of fergusoninid flies and the nematode worms associated with them.

flies have a close mutualistic relationship with nematode worms of the genus *Fergusobia*. In Australia both live together in galls on plant species in the family Myrtaceae. It is unclear whether the flies or the worms are the primary gall-inducers but what is certain is that neither species can survive without the other.

When a female fergusoninid fly uses its sharp ovipositor to lay eggs deep into plant tissue it also deposits young nematode worms. Gall development is stimulated by the feeding of the worms, aided possibly by compounds in the oviposition fluid or the egg of the fly. After the fly egg hatches, the fly larva feeds on the nutritive tissue that forms, and passes through three larval stages before pupating in the gall. While the fly is developing, the young worms also feed on plant tissue and grow into adult females that are capable of reproducing without males. These females lay eggs

The adult female fergusoninid fly *Fergusonina giblindavisi* has a strong ovipositor. Photo: Gary Taylor.

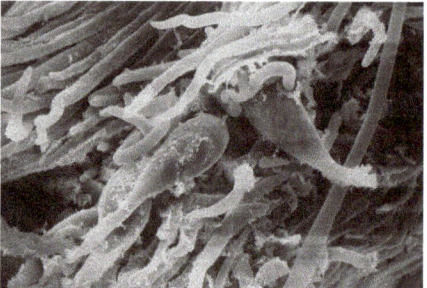

Scanning electron micrograph of eggs (the droplet-like structures) of a *Fergusonina* sp. fly and associated juvenile nematodes, *Fergusobia* sp., in a *Melaleuca* sp. flower bud. Magnified about 350x. Photo: Jim Plaskowitz, USDA Agricultural Research Service, Bugwood.org.

A gall induced by the fergusoninid fly *Fergusonina giblindavisi* and its associated nematode worm species on a flower bud of *Corymbia ptychocarpa*. Photo: Gary Taylor.

within the gall cavity that hatch and become male and female worms. Once adult, these worms mate in the gall and the fertilised females enter the body of the fly just before it turns into a pupa. Worms that enter male flies, or that fail to enter a fly, die. Some time between pupation and emergence of the adult female fly, the now parasitic female worms lay eggs into the circulatory fluid (haemolymph) of the fly. When the worm eggs hatch the young worms migrate to the oviducts of the fly ready to be deposited with its eggs and start the cycle again. Thus the fly needs the worm so that a gall, which supplies the fly with food and shelter, can form – and the worm needs the fly to provide transport to a new gall site.

Enemies of gall-inducing insects

The tough walls and digestion-inhibiting chemicals often secreted in the outer layers of plant galls make it harder for other organisms to compete with, damage, or eat, the gall-inducing insects inside. In spite of this there are still many organisms that have found ways to invade galls for their own purposes and destroy, or disadvantage, the gall-inducers. These enemies of Australian gall-inducing insects include predators, parasitoids, kleptoparasites, inquilines and micro-organisms.

Predators

Predatory insects, such as earwigs and ants, and birds, such as noisy miners, *Manorina melanocephala* and red wattlebirds,

Anthochaera carunculata, have been observed extracting gall-inducing insects from less enclosed Australian galls, such as psyllid pit galls. The caterpillars of some small moths are known to enter partially enclosed thrips galls via their natural openings. These predatory caterpillars apparently chew on the antennae of adult thrips they find inside the gall.

The best known predators of insects inhabiting the harder-to-open, more enclosed kinds of Australian galls are parrots. These birds have strong bills primarily adapted for crushing seeds, and such bills can also be put to good use cracking open galls. Evidence of parrot predation on galls comes from dissections

A swift parrot, *Lathamus discolor*, feeding on a gall in the foliage of *Eucalyptus tricarpa*. Photo: Chris Tzaros.

of accidentally killed swift parrots, *Lathamus discolor*. The gut contents of these birds sometimes include the pupae of gall-inducing fergusoninid flies. It also seems likely that swift parrots, when they are on the Australian mainland during their non-breeding period, rely on gall-inducing psyllids (together with psyllid species that live under lerps) for about half their dietary needs. Other evidence of parrot predation is seen in the characteristic parrot damage on beetle-induced stem galls that have been chewed open and the developing beetle removed and in reports of little corellas, *Cacatua sanguine*, feeding on wasp larvae extracted from galls on acacias.

Invertebrate predators of enclosed Australian galls have also been recorded. For example, mealybugs attack gall-inducing scale insects such as *Cylindrococcus casuarinae,* and the larvae of several wasp species prey on developing fergusoninid fly larvae within their galls. The predatory wasp larvae tunnel from one gall chamber to another searching out the fly larvae and eating them. Larvae of some moth species attack fergusoninid larvae in a similar manner but may also consume everything inside the gall except the outer shell of the gall itself.

Parasitoids

Parasitoids are organisms that live on, or in, other organisms (hosts). They resemble parasites except that, while parasites and their hosts normally co-exist, parasitoids ultimately kill their hosts. Many parasitoids are specialised to survive on only one, or a few, closely related host species.

The free-living adult female parasitoid lays its eggs in, or on, the gall-inducing insect. It achieves this by either inserting its ovipositor through an existing opening in the gall or, if the gall is completely enclosed, by using its ovipositor to pierce the gall wall. When a parasitoid larva emerges from its egg it feeds on the body of its developing host. Eventually the gall-inducer dies. The parasitoid pupates either within the body of

The gall-inducing scale insect *Cylindrococcus casuarinae* on *Allocasuarina verticillata* being attacked by the mealybug, *Sphaerococcus casuarinae*. The mealybugs are the white objects at the base of the gall. Photo: Penny Gullan.

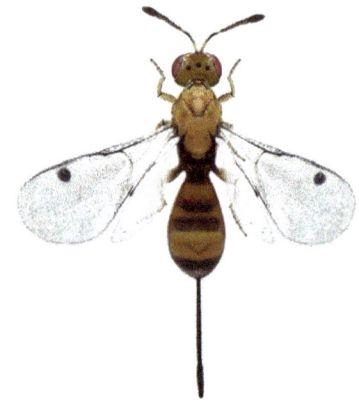

The chalcidoid wasp *Megastigmus* sp. is a parasitoid of the gall-inducing chalcidoid wasp *Leptocybe invasa*, which induces galls on numerous eucalypt species. Photo: Zvi Mendel and Alex Protasov.

the gall-inducer or in the gall chamber, and leaves the gall when it is an adult.

While in the gall the first parasitoid species may be parasitised by another parasitoid species – a hyperparasitoid – and there may also be additional insect species present that are parasitoids of other insect species that have invaded the gall. Thus a large suite of different insect species can emerge from galls that were all originally initiated by a single gall-inducing insect species.

Wasps, especially chalcidoid species, are the main parasitoids that attack Australian gall-inducing insects. Eleven species of parasitoid wasp have been reared from

This chalcidoid wasp, *Sycoscapter* sp., is a parasitoid of fig wasps. Photo: Jean-Yves Rasplus.

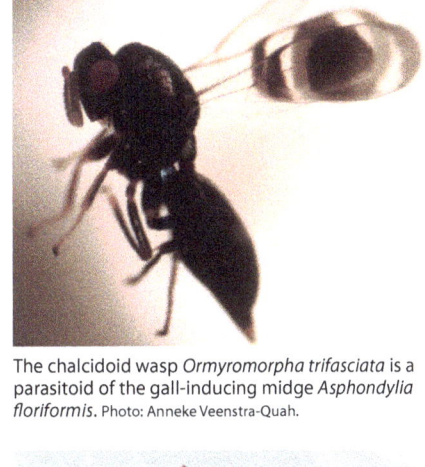

The chalcidoid wasp *Ormyromorpha trifasciata* is a parasitoid of the gall-inducing midge *Asphondylia floriformis*. Photo: Anneke Veenstra-Quah.

This chalcidoid wasp, *Cameronella* sp., is a parasitoid of the scale insect *Apiomorpha sessilis*. Photo: Andy Wang.

The chalcidoid wasp *Cameronella* sp. is a parasitoid of the scale insect *Apiomorpha pharetrata*. Photo: Andy Wang.

banksia stem galls induced by one species of braconid wasp; at least 10 wasp species believed to be parasitoids of fergusoninid flies are known; and wasp parasitoids have been recorded from galls induced by midges, psyllids, scale insects, beetles and other chalcidoid wasps. An example of a parasitoid that is not a wasp is a fly species (family Tachinidae) that occurs in eucalyptus leaf galls induced by a psyllid species.

Other insects that invade galls

Galls are invaded by many kinds of insects and these insects exploit the galls in many different ways. One common invasive strategy is to occupy the gall as an inquiline. An inquiline obtains shelter and food from a gall without actively destroying the gall-inducing species, although it may disadvantage the gall-inducer to some extent. Some inquilines belong to the same insect group as the gall-inducing insect. For example, two inquiline psyllid species are known that build their bi-valve lerps within the gall cavity of a third psyllid species, and there are midge species that specialise in lodging within the galls of other midge species.

Another invasive strategy is to actively kill the original inhabitant and take up

These three wasp species are associated with a fergusoninid fly gall. Photo: Gary Taylor.

residence in its gall. Insects that do this are called kleptoparasites. This name is derived from the Greek word *kleptes,* meaning thief. The best studied examples of Australian insects that steal galls are certain thrips species that specialise in taking over the galls of other thrips species. Fly larvae, moth larvae and ants can also be kleptoparasites of galls.

Of course numerous organisms are likely to invade a gall once the gall-inducing insect has emerged. The empty gall can be a place of shelter for other insects, such as beetles or thrips, and other invertebrates such as spiders. It can also be a source of food for plant-feeding insects and micro-organisms such as fungi.

Micro-organisms

Very little research has been carried out on diseases of Australian gall-inducing insects but there is evidence that disease can potentially be a problem for theses insects, especially when there are numerous individuals living within the same gall chamber. The evidence comes from the relatively recent discovery that some gall-inducing thrips secrete defensive chemicals that inhibit the growth of a well-known pathogenic bacterium, *Staphylococcus aureus.*

If a gall is infested with a fungus it is tempting to simply assume that the fungus killed the gall-inducing insect. This may be true in some cases but the presence of fungal growth within a gall is not conclusive evidence that the fungus was the cause of death. For example, if the gall was caused by a midge, the fungus could have been the insect's food source. If the gall is old the fungus may have invaded after the gall-inducing insect died from other causes or vacated the gall.

Defence against invaders

Gall-inducing insects use a range of chemical, developmental, behavioural and physical mechanisms to help safeguard the gall from potential invaders. The defence mechanisms employed vary according to the species of gall-inducing insect, the morphology of its gall and the type of invader.

Some of the chemical defences have already been mentioned. These include the concentration of hard-to-digest plant secondary compounds in the outer wall of the gall (thought to deter fungi or some organisms that would otherwise chew through the gall wall) and the recently discovered anti-microbial secretions of thrips. Another is the production of waxes. In most scale insect galls a waxy secretion is produced by glands on the outer surface of the insect's body. The waxy secretion – a mix of true waxes combined with other substances like lipids and resins – provides a powdery protective layer that helps prevent contamination of the insect's body and gall by honeydew excreta. Keeping free of honeydew may reduce the incidence of attack by enemies such as pathogenic bacteria and fungi.

The mutualistic relationship between gall-inducing scale insects and honeydew-removing ants is an example of a defence that makes use of the services of another species. The scale insects avoid a build-up of honeydew around the gall opening and the ants are rewarded with food.

The close relationship between gall-inducing midges and fungi is another example of one species gaining protection from the presence of another. Developing midge larvae use fungi for food but at the same time the fungi may be contributing to the defence of the midge larvae. The thick

A gall induced by the scale insect *Apiomorpha attenuate* on *Eucalyptus camaldulensis*. The powdery white substance around the gall opening is a waxy secretion of the insect that helps keep the gall and insect free of honeydew. Photo: Penny Gullan.

mats of fungi in some midge galls are thought to offer protection against parasitoids by making it harder for them to penetrate the gall or locate the place to lay eggs within the gall.

In addition, some gall-inducing midges in other countries limit their exposure to parasitoids by having arrested development known as diapause. Such behaviour has not been recorded for Australian gall-inducing midges but is likely to occur here. The midge larvae grow extremely slowly during the warmer times of the year when parasitoids are most active. During this period the midge larvae are too small to be attacked. By the time they are large enough to be of interest to parasitoids the weather is colder and the parasitoids are less active so attack on the midge galls is less likely too.

As well as attacking individuals of the same species when competing for gall sites, a thrips that initiates a gall – the founder of a gall – will attack kleptoparasitic thrips species that attempt to steal the gall once it is established. In addition, in some thrips species the gall-initiating female produces only fully winged adults, capable of dispersing and inducing new galls, in the second generation of offspring. The first generation produced is composed of individuals that develop into males and females with enlarged forelegs and reduced wings and antennae. These specialised thrips are called soldiers because they actively defend the gall against invaders. They fight by using the ends of their forelegs to stab the membranes behind the front part of the thorax of invading thrips. Gall-inducing thrips that have soldiers produce relatively small broods of young and initiate relatively small, long-lived galls that house two full or partial generations within the gall. Their galls can remain active for up to a year.

Other thrips species do not have soldier cohorts to defend them and employ a different strategy to protect themselves from kleptoparasites. These gall-inducing thrips have relatively big, short-lived galls and produce extremely large numbers of offspring that disperse away from the gall while they are still young larvae. This means there is only a short time available for

Gall-inducing female thrips, *Kladothrips morrisi*. The founding female is on the right and the first generation soldier is on the left. Photo: Laurence Mound.

invaders to attack and kill the gall inhabitants.

It is interesting to note that, because female thrips soldiers have reduced ability to produce young, those gall-inducing thrips species that have soldiers can be regarded as truly social insects. Truly social species – known as eusocial species – have comparatively sterile workers, or soldiers, that care for the reproductive members of the community. Other non-gall-inducing groups that fit this description are ants, termites and social bees and wasps.

The thicker and tougher the wall of a gall, the longer and stronger the ovipositor of a parasitoid needs to be to penetrate the gall and deposit eggs. Over evolutionary time scales, genetic changes (mutations) that produced new gall-inducing species, whose galls had thicker, tougher walls, may have provided at least a temporary respite from the parasitoids adapted to the thinner, softer walls of the ancestral galls. In a similar way, mutations which allowed new species of gall-inducing insects to alter the shape of the gall, or shift to new host plant species, may have provided such species with an enemy-free 'window' for a while. Enemies adapted to searching for specific gall shapes, or on specific plant hosts, would be less likely to find the new gall-inducing species.

The kind of protection afforded by making it harder to penetrate the gall, altering its shape, or changing to a new host plant species, is not likely to be permanent. Eventually new enemies evolve, often from species closely related to the original enemy species. In fact this seems to be true of all defences!

Problems caused by gall-inducing insects

When galls are present in low numbers on native plants their presence can usually be tolerated and may not even be noticed. At best, some people can even enjoy their presence as part of the natural world. However, problems arise when the abundance of galls reaches a stage where they are judged unsightly, or reduce the health and productivity of the plant host.

Australian gall-inducing insects, mainly midges, psyllids and chalcidoid wasps, currently cause losses in home gardens and in the horticultural, agricultural and forestry industries both within Australia and in other countries

Leaf galls on *Banksia coccinea* induced by the midge *Dasineura banksiae*. Photo: Peter Kolesik.

where Australian native plants are cultivated. Gall-inducing insects also have the potential to add to the pressures affecting endangered species.

The scarlet banksia, *Banksia coccinea*, is highly prized because of its large red flowers and attractive foliage. In natural stands of scarlet banksia a gall-inducing midge, *Dasineura banksiae*, causes white globular hairy galls on the underside of leaves. There can be up to 15 galls on a single leaf. The galls are 5–7 mm in diameter and hairy inside and out. Young galls may have a reddish hue. Infestation of commercial scarlet banksia plantations by the gall-inducing midge makes flower stems unsuitable for the fresh flower market. Although infested leaves can be manually removed and the flower stems sold as second grade to the dried flower market, financial returns are reduced.

One Australian gall-inducing chalcidoid wasp, *Bruchophagus fellis*, has expanded its range by utilising not just its natural native host finger-lime, *Citrus australasica*, but also introduced citrus species such as lemon, orange and grapefruit. The wasp induces woody stem galls on young flush growth in spring. Infestations of this insect do not kill citrus trees but repeated attacks can weaken them and cause branch dieback and reduced crop yields. These wasps are a problem for both home gardeners and commercial orchardists in Australia.

Gall-inducing chalcidoid wasps, *Bruchophagus fellis*, on the stem of a citrus tree.
Photo: Dan Papacek, Bugs for Bugs.

The stem gall of the chalcidoid wasp *Bruchophagus fellis* on citrus. Photo: Dan Papacek, Bugs for Bugs.

Problems in overseas countries

The Australian native magenta cherry, *Syzygium paniculatum* (previously known as eugenia) is a common ornamental tree or shrub in California, USA. It is attacked there by an Australian psyllid, *Trioza eugeniae*, which induces pit galls on new leaves and shoots. As well as distorting foliage and stems, the galls inhibit new shoot formation. Sustained high infestations can cause severe weakening of plants, poor growth characteristics and lowered economic value. In addition, the psyllid nymph excretes sugary honeydew that promotes the growth of a black sooty mould on plants. This does not seriously damage the plants but spoils their appearance.

Australian eucalypts are grown worldwide as a source of wood for fuel, pulp for paper-making, nectar for honey and to extract eucalyptus oil. They are also planted as ornamentals or to provide wind-breaks. Places where eucalypts are now grown include New Zealand, India, China, Thailand, Israel, Italy, Spain, Kenya, Uganda, Ethiopia, South Africa, Turkey, the United States, Colombia, Brazil, Bolivia, Chile, Argentina and Peru. There are several gall-inducing Australian chalcidoid wasp species that have become serious pests of these introduced eucalypts.

One Australian chalcidoid wasp, *Epichrysocharis burwelli*, causes small, reddish or brownish, blister-like galls on the leaves of native lemon-scented gum, *Corymbia citriodora*. There can be more than 40 galls per cm² of leaf surface. In California and Hawaii such high concentrations of galls decrease the plant's value to the nursery trade. In Brazil the galls cause premature leaf fall that reduces the

The magenta cherry, *Syzygium paniculatum*, is a native Australian plant grown as an ornamental in California, USA. Photo: Bart Wursten.

amount and quality of essential oils that can be extracted from the eucalypt.

Another chalcidoid wasp, *Selitrichodes globulus*, causes multi-chambered stem galls on southern blue gum, *Eucalyptus globulus*, in California. There can be up to 20 wasps per 5 mm of branch. The galls disfigure the tree and reduce its vigour. This wasp has only recently been discovered but it poses a potentially serious problem if it spreads because its host, blue gum, is widely planted in commercial plantations.

The chalcidoid wasp *Ophelimus maskelli*, which is already a serious pest of eucalypt plantations, induces single-celled galls on the leaves of at least 14 eucalypt species in the Mediterranean region. This

Pit galls on leaves of magenta cherry induced by the psyllid *Trioza eugeniae*. Photo: Whitney Cranshaw, Colorado State University, Bugwood.org.

Masses of stem galls of the chalcidoid wasp *Selitrichodes globulus* on southern blue gum, *Eucalyptus globulus*.
Photo: Gevork Arakelian.

pest is of particular concern because it severely damages river red gum, *Eucalyptus camaldulensis*. River red gum is the most widespread introduced hardwood species in the Mediterranean and in the Middle East. Heavy infestations of up to 36 galls per cm² of leaf cause premature leaf drop.

Another gall-inducing chalcidoid pest, *Leptocybe invasa*, was first recognised in young eucalypt plantations in the Middle East, the Mediterranean and northern and eastern Africa. It has now spread to South Africa, India and Brazil. Most recently it has been detected in Florida, USA. This wasp causes galls on stems, leaf midribs and petioles of numerous eucalypt species, including river red gum. There can be more than 50 galls per leaf. Heavy concentrations of galls like this prevent further development of infested growth.

Galls of the chalcidoid wasp *Ophelimus maskelli* on leaves of *Eucalyptus camaldulensis*.
Photo: Zvi Mendel and Alex Protasov.

The chalcidoid wasp *Oncastichus goughi*, which galls Geraldton wax, *Chamelaucium uncinatum*, is considered to be a pest in Australia, Israel and California, USA. Recently the wasp was also found in commercial stands in Peru. The wasp induces galls on the new stems and needle-like leaves of the plant. The galls are only small, up to 6 mm long and less than 2 mm wide, but their presence results in severely deformed branching patterns that reduce the plant's suitability for the cut flower and nursery industries.

At least three Australian chalcidoid wasp species, *Quadrastichodella nova*, *Moona spermophaga* and *Leprosa milga*, induce galls in eucalypt seed capsules. This poses quarantine problems for exporters of eucalypt seed. *M. spermophaga* can be dormant in seeds for at least 10 years and can survive carbon dioxide fumigation procedures that normally kill other insect pests. Seed-galling chalcidoid wasps are now known to have spread to the USA, South Africa, Italy, Argentina, Israel, Turkey and Spain.

Adverse effects on endangered species

The way in which gall-inducing insects can disadvantage endangered species is illustrated by the competition between two gall-inducing midge species, *Asphondylia floriformis* and *A. sarcocorniae,* and the critically endangered orange-bellied

Stem galls induced by the chalcidoid wasp *Leptocybe invasa* on a eucalypt. Photo: Zvi Mendel and Alex Protasov.

Geraldton wax, *Chamelaucium uncinatum*, a native plant grown for its attractive flowers and foliage. Photo: Forest and Kim Starr.

parrot, *Neophema chrysogaster*. In 2000 there were only about 180 of these birds left in the wild. Each year the parrot migrates to mainland Australia from Tasmania to overwinter in Victorian coastal salt marshes. Once there it relies predominantly on the seeds of the beaded glasswort, *Sarcocornia quinqueflora*, for food. The midges also rely on this plant to initiate their galls. Gall formation prevents the development of normal shoots and decreases seed production.

Of course, the midges and the parrot have co-existed for a long time, so competition with the midges is not the primary cause of the parrot's decline. The main reason the parrot is endangered is that its habitat, and consequently its food source,

has been shrinking due to residential, industrial and recreational development along the Victorian coastline. However, if large outbreaks of the midges occurred they would further reduce the seeds available for the parrots to eat and could contribute to the decline in numbers of the parrots.

Another example of the potential impact of a gall-inducing insect on an endangered species comes from Uganda. Australian eucalypts are grown in rural areas there to provide firewood for domestic use so that people can stop cutting down rainforest trees for this purpose. One reason that it is important to conserve these rainforests is that they are the habitat of endangered gorillas. Galls caused by an Australian chalcidoid wasp on the leaves of the introduced eucalypts are damaging the trees to such an extent that people are once more turning to the rainforest for firewood. Consequently the threat to the survival of the gorillas is increasing.

The source of food for the orange-bellied parrot may be reduced by the gall-inducing midges *Asphondylia floriformis* and *A. sarcocorniae*. Photo: Ian Montgomery.

Flower-like galls induced by the midge *Asphondylia floriformis* on the leaves of beaded glasswort, *Sarcocornia quinqueflora*. Photo: Anneke Veenstra-Quah.

Stem galls on beaded glasswort, *Sarcocornia quinqueflora,* induced by the midge *Asphondylia sarcocorniae*. The galls look like swollen segments of the horizontal stem. The thinner stems show the normal form of new stem growth. Photo: Anneke Veenstra-Quah.

When gall-inducing insects become pests

Unfortunately, the enemies of gall-inducing insects are often left behind when gall-inducing insects are accidentally transported outside their natural range on their host plants. This can allow the numbers of the gall-inducing insect to rise dramatically and cause plants to suffer more damage.

There can also be problems even when the enemies of the pest are introduced, if the environmental conditions in a new habitat favour the pest rather than its enemies. The timing of pesticide applications and the kind of pesticides used are additional factors that can reduce the abundance of the enemies of a gall-inducing insect more than the numbers of the pest itself.

If gall-susceptible host plants are pruned or fertilised at a time that stimulates them to re-sprout when the gall-inducing species that attack them are looking for young plant tissue to lay eggs in, or start galls, increased pest abundance can result due to the increased number of sites available for gall initiation. In a similar way, planting large monocultures of gall-susceptible plant species, such as single-species eucalypt plantations, encourages pest outbreaks by improving the chances that a dispersing pest insect will find a suitable host.

Apart from growing plant species or varieties which are known to be resistant to attack, the problems associated with gall-inducing insects can be minimised by improving the likelihood that natural enemies, which can help keep the pests in check, are present. One way to do this is to cultivate plant species that occur naturally in the growing area. This increases the chance that the enemies of gall-inducing pests attacking the plant will also be present.

Biocontrol – that is, introducing natural enemies of gall-inducing pests to gardens, orchards or plantations – is another way of dealing with pest outbreaks. The enemies

This chalcidoid wasp, *Closterocerus* sp., is a parasitoid of the gall-inducing chalcidoid wasp *Ophelimus maskelli*, which is a pest of eucalypt plantations. The parasitoid is thus a potential biocontrol agent. Photo: Zvi Mendel and Alex Protasov.

chosen are usually parasitoids that specifically target the pest species. Finding the right parasitoids can be costly and time consuming, but this approach is worthwhile because it often results in the most effective and long-term control.

If natural enemies of the pest are known to be present it may be best to protect them by not using pesticides. Alternatively, suitable pesticides can sometimes be applied to plants at times when gall-inducing pests, but not their enemies, are most susceptible.

Sometimes it is feasible to reduce the size of future attacks by simply cutting off infested branches and destroying them. In this case it is important to either burn the branches or wrap them securely in plastic and place them in the garbage, because their load of developing pest insects may still be viable.

Another method is to hang sticky traps among infested plants to catch gall-inducing insects as they disperse from their galls. In addition, pruning and fertilising plants can be timed so that they do not stimulate sprouting when it would favour gall-inducing insects.

Usually a combination of appropriate pest management strategies is needed to reduce gall-inducing pest numbers to acceptable levels. As with all pests problems the complete eradication of the pest species is rarely achievable.

Benefits associated with gall-inducing insects

Several Australian gall-inducing species play a vital role in helping control Australian plants that have become weeds. In addition, Australian gall-inducing wasps are essential pollinators of native figs and some gall-inducing insects, and their galls, are still highly valued by Australian Aboriginal people as traditional food. Naturally occurring benefits like these are often called ecosystem services.

Acacias in South Africa

There are at least 13 Australian species of *Acacia* that have become invasive weeds in South Africa. They can form dense thickets that impede agriculture, suppress native plants and animals, increase fire risks and interfere with flow rates of rivers and streams. Acacias were originally introduced into South Africa as ornamental plants or shade trees and to reclaim or stabilise sand dunes. Other species were, and still are, important sources of firewood, timber for furniture, high-grade pulp wood and tannin.

One of these invasive acacias, long-leaved wattle, *Acacia longifolia*, was introduced into South Africa in 1827 to stabilise dunes and for shade and ornamental purposes. The concentration of seeds dropped in the soil by this plant can be as high as 34 000 seeds per m². Seeds germinate in enormous numbers after fire. An Australian gall-inducing chalcidoid wasp, *Trichilogaster acaciaelongifoliae*, was released in 1982–83 to reduce the damage caused by this acacia. It was the first time

that a wasp had been used as a biocontrol agent.

Trichilogaster acaciaelongifoliae induces galls in young flower buds of long-leaved wattle. The multi-chambered galls prevent development of flowers and can reduce seed production by more than 95 per cent. High concentrations of galls cause breakage and mortality of large branches and stems. The wasp has proven to be a highly effective natural enemy of this invasive acacia and has spread to all the areas in South Africa where the acacia grows.

Galls of the chalcidoid wasp *Trichilogaster acaciaelongifoliae*, which was introduced into South Africa to control invasive Australian long-leaved wattle, *Acacia longifolia*. Photo: John Hoffmann.

Galls of the chalcidoid wasp *Trichilogaster signiventris*, which was introduced into South Africa to control invasive Australian golden wattle, *Acacia pycnantha*. Photo: John Hoffmann.

A second Australian gall-inducing chalcidoid wasp species, *Trichilogaster signiventris*, was introduced into South Africa in 1992. This wasp was selected to combat golden wattle, *Acacia pycnantha*. Golden wattle was brought into South Africa in the mid 1800s for dune reclamation. It was also planted as an ornamental. Although dense thickets of golden wattle are uncommon, and the acacia is relatively sparsely distributed in two widely separated areas in South Africa, it is still considered an invasive threat.

Trichilogaster signiventris targets mainly flower buds but also attacks non-reproductive plant parts, such as stem tips. The average weight of a mature gall is 907 mg. The galls virtually eliminate seed production in golden wattle and the high densities of such heavy galls often cause branches to break from trees or their main trunks to split. To date it appears that this gall-inducing wasp will be as effective at helping control golden wattle in South Africa as *T. acaciaelongifoliae* has been in the management of long-leaved wattle.

Biocontrol of long-leaved wattle and golden wattle does not rely solely on gall-inducing wasps. An Australian seed-eating weevil, *Melanterius ventralis*, is also involved in the control of long-leaved wattle and gall-inducing Australian midges (*Dasineura* and *Asphondylia* spp.) are currently under consideration as additional biocontrol agents for these and other invasive acacia species.

Gall-inducing Australian midges are good candidates for the biocontrol of other invasive acacias in South Africa, such as *Acacia mearnsii*, *A. melanoxylon*, *A. dealbata* and *A. decurrens*, which also have commercial significance. It is important that control methods applied to such species do not interfere with the production of tannin, timber and pulpwood products. Australian gall-inducing midges are likely to be suitable biocontrol agents in these situations because their galls are tiny (2–4 mm wide). They have an impact on seed production only and do not damage vegetative plant parts, as do the heavy masses of galls induced by chalcidoid wasps on long-leaved and golden wattles. Two gall-inducing midge species, *Dasineura dielsi* and *D. rubiformis*, have already been released.

Paperbark trees in Florida

Another invasive pest is the paperbark tree, *Melaleuca quinquenervia*. It was introduced into Florida, USA, in 1886 as an ornamental plant and as a forestry crop to grow on the edge of the Everglades. Later it was planted to lower soil moisture in parts of the Everglades destined for housing and industrial developments. The tree produces massive amounts of seed and can form dense forests up to 30 m high.

Paperbark forests have now invaded about 610 000 ha of southern Florida. These forests have displaced native plants and animals, increased the chance of fire and have reduced fishing, hunting and air-boating activities. The trees are also thought to cause allergic skin and respiratory reactions in susceptible people. The corky bark of the paperbark has meant that it is

A dense stand of the invasive paperbark tree *Melaleuca quinquenervia* in Florida, USA. Photo: Forest and Kim Starr.

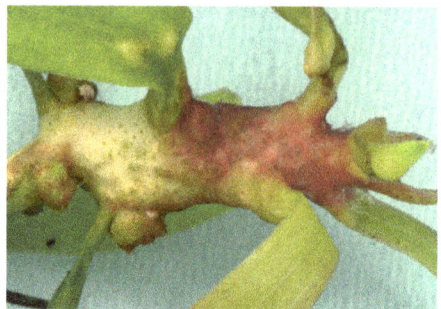

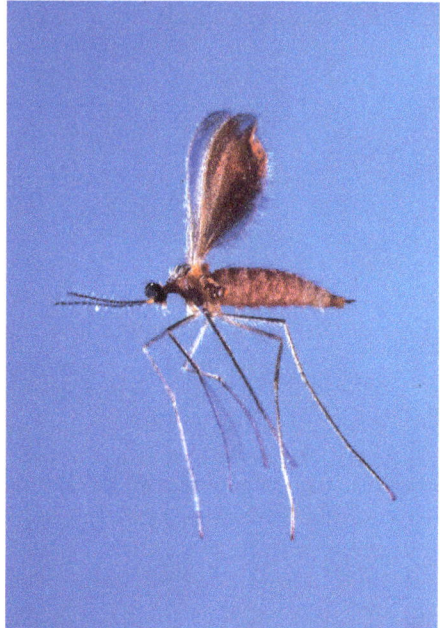

A gall induced by the midge *Lophodiplosis trifida* on a soft developing shoot of the paperbark *Melaleuca quinquenervia* and (bottom) cut open to reveal the gall chambers and developing midges. Photos: Susan A Wright, USDA, ARS, Invasive Plant Research Laboratory.

An adult midge, *Lophodiplosis trifida*, a possible biocontrol agent for the invasive paperbark *Melaleuca quinquenervia*, in Florida, USA. Photo: Susan A Wright, USDA, ARS, Invasive Plant Research Laboratory.

not suitable for forestry purposes. Plans to restore the damaged wetlands and prevent further spread of paperbark trees involve cutting down trees, spraying them with herbicides and introducing biocontrol agents, including gall-inducing insects.

An Australian fergusoninid fly/nematode complex, *Fergusonina turneri/Fergusobia quinquenerviae*, reduces seed production of the paperbark tree by inducing flower bud galls that suppress bud development. This gall-inducing complex was released in Florida but is proving difficult to establish. Another Australian gall-inducer, the midge *Lophodiplosis trifida* has since been released and seems to be establishing well. The midge initiates stem galls that impede plant growth and kill young paperbark trees.

Coastal tea tree in Australia and South Africa

Coastal tea tree, *Leptospermum laevigatum*, has become an invasive weed both within Australia, in areas outside its natural south-eastern coastal range, and in South Africa. This plant forms dense thickets that can shade out other plants or deprive them of water. It is especially considered a problem in Western Australia where it has been planted extensively to stabilise sand dunes.

Two Australian gall-inducing midge species, *Dasineura strobilas* and *D.*

tomentose, have recently been described. They initiate galls in developing stems and flower buds of coastal tea tree causing disruption to plant growth and preventing fruit and seed set. The two midges have potential to be biocontrol agents for this invasive weed.

Gall-inducing insects that pollinate figs

The story of figs and the numerous wasps species associated with them (including pollinating wasps) is a complex one.

The flower clusters (inflorescences) of fig trees are completely concealed. Hundreds of tiny flowers line the inside cavity of each receptacle (syconium) that will become the fig fruit. Each syconium has only one extremely small opening (ostiole) at its exposed end.

The pollinators of fig flowers are tiny (1–2 mm long) gall-inducing chalcidoid wasps of the family Agaonidae. Some pollinating wasps induce galls in only one fig species but others are not completely host specific. For example, the Australian wasps *Pleistodontes froggatti* and *P. imperialis* both induce galls on Moreton Bay fig, *Ficus macrophylla,* and on Port Jackson fig, *F. rubiginosa.*

Gall-inducing fig wasps exhibit strong sexual dimorphism that reflects the different activities of males and females. The role of male wasps is played out mainly inside the developing fig fruit. They chew open galls of female wasps, fertilise the females and then burrow through the fig wall to allow the females to escape. Consequently, most male fig wasps lack wings and their eyes, antennae and the end segments of their legs

The Australian native Moreton Bay fig, *Ficus macrophylla.* Photo: Forest and Kim Starr.

A male fig wasp, *Pleistodontes froggatti*, inside a fig of the Moreton Bay fig. Photo: Jean-Yves Rasplus.

A female fig wasp, *Pleistodontes addicotti*, on its host fig, *Ficus crassipes*. Photo: Jean-Yves Rasplus.

are reduced. Other parts of their legs are enlarged or strengthened to assist with digging the exit hole in the fig wall.

Female fig wasps have relatively large wings and functional eyes that facilitate dispersal to colonise new host figs. The heads and antennae of female wasps are modified and their bodies are flattened to enable them to enter the syconium through the ostiole. Once inside a new syconium, the female fig wasps lay eggs by inserting their ovipositors into the tubes (known as styles) that lead into the ovaries of the flowers. At the same time, the female wasps deposit pollen, acquired from the figs they developed in, on an area at the top of the style (stigma).

Pollination has been studied in only a few Australian fig tree species. They can be monoecious or dioecious.

In monoecious fig trees each inflorescence has some female flowers in which wasps induce galls and other female flowers that are not galled. The first kind of flower results in the production of wasps and the second seeds. Male flowers develop later than female flowers within the same syconium and produce pollen that will be

transported by the next generation of mated females when they leave their natal fig to oviposit in a newly developing syconium.

Dioecious figs have 'functionally male' trees and female trees. Female wasps are attracted to syconia on both kinds of trees. The male trees have female flowers with short styles, in which the wasps can reproduce, and male flowers for pollen production. The female trees only have female flowers and the styles of these flowers are too long for the ovipositors of the wasps to reach the flower ovules. Female wasps that enter syconia on female trees deposit pollen (from male trees) that fertilises the flowers but no galls are initiated. Galls and wasps only develop in female flowers on male trees. Only seeds are produced from female trees.

The association of figs and their pollinating wasps is thought to date from some time in the Cretaceous 90–70 million years ago. It is another example of mutualism. The wasps are completely dependent on the fig for development of their larvae and the figs need the wasps to pollinate them so that seed can be produced. Neither can reproduce without the other.

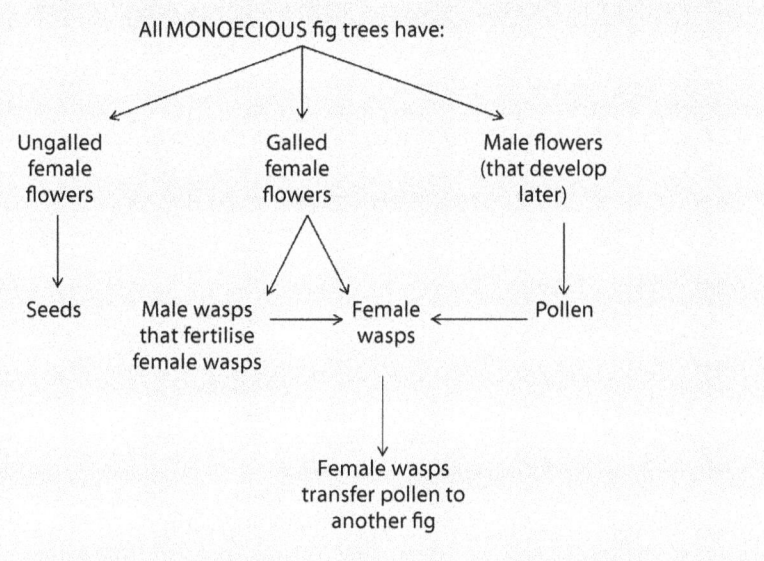

All MONOECIOUS fig trees have:

Ungalled female flowers → Seeds

Galled female flowers → Male wasps that fertilise female wasps → Female wasps ← Pollen ← Male flowers (that develop later)

Female wasps transfer pollen to another fig

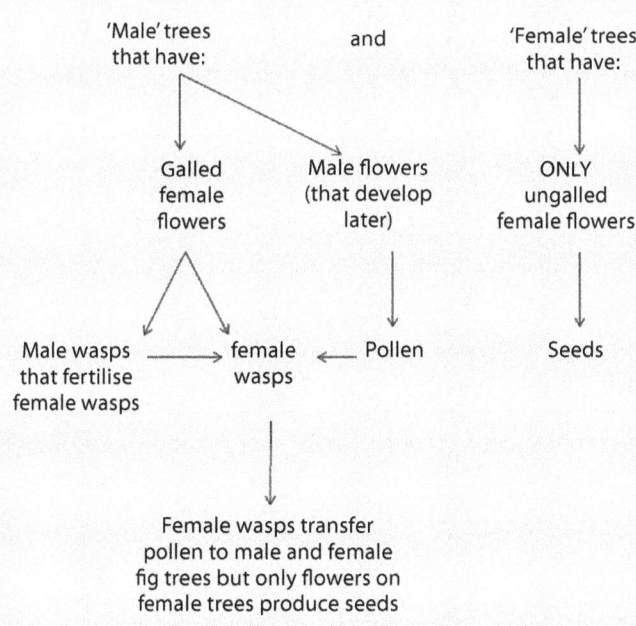

All DIOECIOUS fig trees have:

'Male' trees that have: and   'Female' trees that have:

Galled female flowers   Male flowers (that develop later)   ONLY ungalled female flowers

Male wasps that fertilise female wasps → female wasps ← Pollen   Seeds

Female wasps transfer pollen to male and female fig trees but only flowers on female trees produce seeds

Gall-inducing insects as bush tucker

There are at least two species of gall-inducing insect that provided food for Australian Aboriginal people in the past and are still eaten today. One is a wasp that induces round, lumpy galls, the size of a large marble, on mulga, *Acacia aneura*. The gall is called a mulga apple. Little is recorded about mulga apples beyond the fact that they have a sweet taste and that both the gall itself, and the wasp larva in its centre, are eaten. The larva is considered to be the best part!

More is known about another gall-inducing insect species used for food by Aboriginal Australians. This is the scale insect, *Cystococcus pomiformis*, which induces galls on *Corymbia* species. The gall is called a bloodwood apple or a bush coconut. It is a woody, round, stem gall that can be up to 9 cm in diameter. The large juicy female scale insect can be more than 2 cm long. The gall is broken open and the female extracted and eaten raw. Then the white nutritious lining of the gall is scraped out and consumed.

This gall of the scale insect *Cystococcus pomiformis* has been sawn open to show the edible lining of the gall and the female insect. The red-brown wood shavings are from the hard outer layers of the gall wall and fell into the gall cavity when it was being cut open. Photo: Emma Woodward.

Studying galls and their insects

In order to answer the numerous questions still surrounding Australian gall-inducing insects and their life in galls, we need to collect and study them. This work need not be restricted to professional scientists but can be carried out by diligent amateurs too. The sedentary nature of the gall-inducing way of life is one of the factors that make gall-inducing insects good candidates for study. At least the host plant of the insects and its geographical location are always obvious. On the other hand, most galls harbour many other species apart from the gall-inducer, so it is often difficult to determine which species initiated the gall. Careful observation and detailed studies are needed to tease out the complex relationships of insect life within galls.

The simplest way to catch gall-inducing insects is to remove their galls from the host plants and then either cut the galls open to extract the insects or store the galls and collect the insects that ultimately emerge. If using the latter method it may be necessary to gather galls at several different times because if the insects are too immature when removed from the host plant they will die without developing into adults. Adult scale insect females that never leave their galls will also die once the gall is no longer attached to the host plant.

Gall-collecting equipment

Unless you are collecting galls in your own garden, the first step is to obtain permission to collect in the area you want to survey.

Collection of native plants and animals (including insects) is not permitted on public lands, such as National Parks, without first obtaining an official collection permit. It is also polite to obtain at least verbal permission from private landowners.

When out in the field collecting galls, especially for long periods in hot dry areas of the Australian bush, it is wise to use sunburn cream and wear sunglasses, a shady hat, a long-sleeved shirt, long trousers and sturdy shoes. It is also important to carry plenty of water to drink. Other basic items needed when collecting and studying galls are listed on page 54.

After recording the date and collection location (as precisely as possible) in your notebook, the galls and host plants can be photographed *in situ*. Specimens of each can then be cut from the plants. Take care

Collecting specimens with a long-handled pruner.
Photo: John Ludwig.

Basic items needed for collecting galls

Item	Use
Official collection permit if required	Allow collection on public lands
Permission from landowner	Allow collection on private property
Map or GPS unit if available	Obtain location co-ordinates
Notebook and pen or pencil	Record data
Camera	Photograph galls in their natural setting
Secateurs, penknife, long-handle pruner	Cut galls from host plants
Paper bags (various sizes)	Hold gall specimens for transport
Adhesive tape	Seal paper bags
Sheets of newspaper (and a plant press if available)	Hold plant specimens
Large box	Hold bagged galls and plant specimens

Basic items needed for processing and studying galls

Item	Use
Plant press or heavy books	Hold plant specimens flat while they dry
Magnifier	View details of specimens
Light source	Aid specimen viewing
Vice or locking pliers	Hold tough-walled galls for dissection
Scalpel or penknife	Dissect galls
Cutting board or dish	Cut galls on
Fine forceps	Extract insects from galls
Camera	Photograph open galls and insects
Ruler or tape measure	Measure insects and galls
Notebook and pen or pencil	Record data
Clean glass jars (e.g. used jam jars)	Store galls for rearing adults
Box of tissues and box of strong rubber bands	Cover jars
Paper labels	Label jars

Examining galls on the host plant. Photo: Ros Blanche.

not to damage the plants any more than necessary.

Each plant specimen is best placed inside a folded sheet of newspaper until it can be pressed flat to dry later. Gall specimens are each placed in separate paper bags. The tops of the paper bags can be folded down and sealed with adhesive tape. Both plant and gall specimens should be labelled on the newspaper sheet and paper bag respectively. The information on these labels should include the collection location, the date (in an unambiguous format e.g. 4 November 2012), the name of the gall-inducing insect or the host plant (if these are known), a brief description of the specimen, the name(s) of the collector(s) and, if collecting from multiple host plant species, a unique number or letter that links the gall specimen to its host plant specimen.

Studying galls

Once the specimens arrive at the home laboratory they need to be processed. Plant specimens can be flattened under heavy books if a plant press is not available. The details of gall specimens, such as their size, colour, shape and location on host plant, can be recorded in your notebook. Galls to be dissected are easy to cut open with a smooth-edged knife if the gall walls are soft. They can be held in place by hand on a cutting board. Galls with tough walls may need to be anchored securely to avoid the knife slipping during dissection. A vice, or a pair of locking pliers, is useful for this purpose and a knife with a serrated edge often works best with tough-walled galls. Fine forceps, such as jeweller's forceps, are best for extracting the insects from the dissected galls.

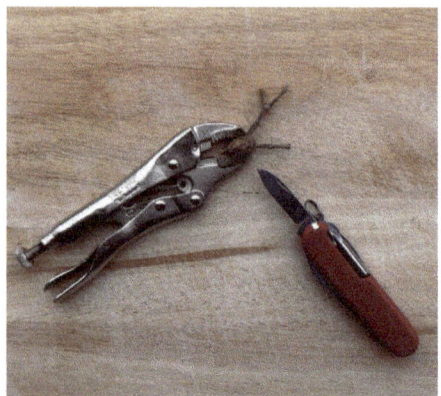

Hand-held locking pliers, a penknife and a cutting board are useful tools when dissecting galls.
Photo: John Ludwig.

Plant and gall specimens collected for study.
Photo: Jason Van.

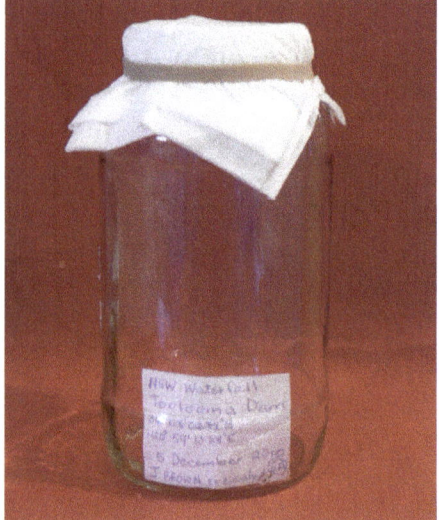

An improvised rearing chamber. Note that the label is placed inside the jar to ensure that it stays associated with the correct gall specimen.
Photo: John Ludwig.

Many of the insects extracted from dissected galls for study and identification will be soft-bodied. The concentrated ethanol that scientists use to preserve soft-bodied specimens like these is not readily available to the general public. Rather than preserve specimens for identification it is easier to simply photograph, or draw, the insects and carefully record their size and appearance.

Some kind of magnifier will be needed in order to see the details of these usually tiny insects. A binocular microscope is best but if not available then a good quality hand-held magnifying glass, or lamp with a built-in magnifier, will allow some of the distinguishing features of the different insect groups to be seen and recorded.

Galls collected for rearing adults can each be stored separately in suitably sized glass jars with the tops covered by a layer of paper tissue secured with a rubber band. This method allows the galls to be easily checked to see when adults emerge. It also prevents moisture build up in the jar that could cause fungal growth to contaminate the galls. Each jar must have a small paper label inside containing the same information that was on the paper bag that the specimen came in. Once the adults emerge from the galls the jars can be placed in a freezer to kill the insects. The galls can then be returned to their original paper bags and retained for possible future reference. Soft-bodied adults that emerge from galls may need to be drawn or photographed for identification but hard-bodied insects can be preserved in a dry collection.

Keeping gall insects in a dry collection

Hard-bodied insects can be pinned, if large enough, but if very tiny, they are glued to the end of small cardboard triangles attached to pins (carding). A small label card is also attached to each pin underneath the insect specimen. This label lists the collection location, the date of collection, the collector(s) name(s) and sometimes the host plant and method of collection.

Preserving insects in a dry collection like this requires specialist equipment such as insect pins, micro-pins, card points, gum Arabic (water-soluble glue) and storage boxes. Many of these items can be purchased from entomological supply companies such as Australian Entomological Supplies <http://www.entosupplies.com.au/>.

The correct techniques for pinning or carding hard-bodied insects and maintaining a dry collection can be learned from basic entomology textbooks and courses, or by following online instructions. The ability to classify insects and their host plants into broad groups can also be acquired this way but identification of specimens to species level will probably require the assistance of a professional entomologist or botanist.

Scientists working in natural history museums, universities, botanical gardens and the Australian National Insect Collection (ANIC) may be able to help. Many scientists have a passion for their subject and are happy to encourage others to share it. However, the scientists' work commitments may not always allow them enough spare time to become involved in your project. Don't be discouraged by this. There are alternative sources of help available! Groups such as Australian Native Plant Society Australia (ANPSA), field naturalists clubs and amateur entomology groups often have knowledgeable members who can assist you with insect and plant identifications.

Managing the information you collect

Ideally all information collected should be entered into a computer spreadsheet, in a consistent and accurate manner, so that it is in a suitable form for possible future analyses. The location and description of the galls, the insects and their host plants is likely to be the first information established. The relevance of other data depends on the questions you would like to answer.

For example, if the aim is to discover whether a particular gall-inducing insect occurs in all areas in which its host plant occurs, then the results from searches for the insect at numerous sites within the natural geographic distribution of the host plant would be needed. Ideally these sites would include the range of environmental conditions that are suitable for the host plant. Information about the environment at the sites (e.g. annual maximum and minimum temperatures, annual rainfall, vegetation and soil types) should also be recorded. These data may provide clues as to why the insect and its host plant do, or do not, have the same geographic distribution.

Depending on the insect/host plant being studied, a project like this could be costly in terms of travel expenses and time. You might want to extract the most value from a study that involves a widespread host plant species by expanding the survey to include all the gall-inducing insect species that you find on that host plant.

A study that could be done closer to home would be to monitor the development of a gall-inducing insect species from the time the gall first appears on the host plant until the insect is mature. Samples of galls collected and dissected over a period of days, or weeks, will reveal the characteristics of the developing insects and their galls. Weather conditions recorded at the collection site on each date (e.g. temperature, rainfall, humidity) may provide useful information about what is influencing insect and gall development rates.

A study carried out in a similar way could focus on the parasitoids and inquilines associated with the gall. Opening a gall is a bit like unwrapping a 'lucky dip' parcel. You never know exactly what you will find inside! Additional data to record at each collection time would be whether or not there was evidence that the gall had been invaded. Signs of invasion include the presence of more than one insect species inside the gall and oviposition damage to the gall wall. At each collection time, as well as dissecting galls, it may also be necessary to store samples of the gall for rearing the non-gall inducers. Emergence dates of adult parasitoids or inquilines should be recorded, the insects described and drawn or photographed, and hard-bodied specimens preserved in a dry collection.

If you are interested in insect behaviour, and lucky enough to see free-living adults mating or ovipositing on their host plant, you might enjoy photographing them and making notes about what they do. Records of the activities of gall-inducing insect larvae or adults, observed at the time they emerge from stored galls, can also provide useful insights into their requirements (e.g. whether pupation occurs in the gall or soil; how long after emergence the adults mate). Observations like these, made in the field or in the home laboratory, can provide natural history information that is lacking for many Australian gall-inducing insects.

Study by scientists

In Australia insect taxonomists are continually discovering new gall-inducing species and formally describing them. Other scientists, such as ecologists, are exploring the environmental factors that affect the number and distribution of gall-inducing insect species that occur in Australia. Molecular biology methods are being used by others to clarify the sequence of events involved in the evolutionary development (phylogeny) of gall-inducing insects and their host plants so that questions about evolutionary relationships can be resolved. Teams of plant and insect specialists are seeking gall-inducing insects with potential to be used as new biocontrol agents for invasive weeds, and evaluating parasitoids that may become new biocontrol agents for gall-inducers that have become pests.

A few years ago a group of primary school children in Canberra, guided by some scientists, discovered and described a unique kind of gall on a small shrub called *Bossiaea grayi*. This plant grows in sand among boulders on the banks of the Murrumbidgee River and its tributaries in the Australian Capital Territory. The seeds

Children and a scientist studying galls in the laboratory. Photo: John La Salle.

Seeds of *Bossiaea grayi*. The dark coloured seed on the right has a normal elaiosome but the elaiosome of the seed on the left is enlarged because it has been galled. Photo: Children and John La Salle.

The small shrub *Bossiaea grayi*. Photo: John Ludwig.

of *B. grayi* are dispersed by ants. To attract the ants each seed has a specialised oil-rich structure (elaiosome) that ants like to eat. The school children found that a previously undescribed chalcidoid wasp induced galls in *B. grayi* elaiosomes. This was a particularly exciting find because it was the first time a gall had ever been discovered on an elaiosome.

With the help of the scientists the children used a digital camera and a microscope to take photographs of the new wasp species and the galled elaiosomes, and a scanning electron microscope and computer software to produce electron micrographs of the insect. They took measurements to compare the new species with ones already described, participated in discussions on how the new species was different from other species and prepared slides for a talk about their work that was presented by one of the scientists at the International Congress of Entomology held in Brisbane in 2004. Undoubtedly they had fun naming the new wasp species. The name they chose was *Tanaostigmodes shrek*, after the much-loved cartoon character. The achievements of these school children should serve as an inspiration for all amateur gall-researchers!

The head of the chalcidoid wasp *Tanaostigmodes shrek* that induces galls on the elaiosomes of *Bossiaea grayi*. Photo: Children and John La Salle.

Further reading

Crespi BJ, Morris DC and Mound LA (2004) *Evolution of Ecological and Behavioural Diversity: Australian Acacia Thrips as Model Organisms.* ABRS and CSIRO, Canberra.

CSIRO Division of Entomology (1991) *The Insects of Australia.* 2 volumes. Melbourne University Press, Melbourne.

Fagan MM (1918) The uses of insect galls. *The American Naturalist* **52**, 155–76.

Hardwick S, Harper M, Houghton G, La Salle A, La Salle S, Mullaney M and La Salle J (2005) The description of a new species of gall-inducing wasp: a learning activity for primary school students. *Australian Journal of Entomology* **44**, 409–414.

Hollis D (2004) *Australian Psylloidea.* Australian Biological Resources Study, Canberra.

Labandeira CC and Phillips TL (1996) A Carboniferous insect gall: insight into early ecologic history of the Holometabola. *Proceedings of the National Academy of Science, USA.* August **93** (16), 8470–8474.

Morgan FD (1984) *Psylloidea of South Australia.* Handbooks Committee, the South Australian Government, Adelaide.

Raman A, Schaefer CW and Withers TM (Eds) (2005) *Biology, Ecology and Evolution of Gall-inducing Arthropods.* 2 volumes. Science Publishers, Enfield, New Hampshire.

Russo R (2006) *Field Guide to Plant Galls of California and other Western States.* University of California Press, Berkeley.

Turnbull C, Hoggard S, Gillings M, Palmer C, Stow A, Beattie D, Briscoe D, Smith S, Wilson P and Beattie A (2010) Antimicrobial strength increases with group size: implications for social evolution. *Biology Letters*, <http://rsbl.royalsocietypublishing.org/>.

van Noort S and Rasplus J-Y (2004) *Figweb. Figs and fig wasps.* Iziko Museums, Cape Town, <http://www.figweb.org/Figs_and_fig_wasps/index.htm>.

Glossary of scientific terms

apterous – having no wings.

biocontrol – the use of natural enemies to reduce the damage caused by a pest population.

bivoltine – having two broods of offspring per year.

carding – glueing tiny insects to small cardboard triangles attached to pins.

cecidology – the study of galls.

chenopod – any plant of the Chenopodiaceae (goosefoot) family, which includes spinach, beetroot and pigweed.

cuticle (of an insect) – the exoskeleton – an extracellular layer covering the outer surface of the insect; it is mainly composed of chitin filaments embedded in a protein matrix, plus smaller amounts of lipids and phenols.

diapause – a period of suspended growth and development.

dioecious – having the male and female reproductive organs borne on separate individuals of the same species.

ecology – the relationship between organisms and their environment.

ecosystem services – the important benefits for human beings, such as clean air and water, that arise from healthy interactions of communities of organisms with their physical environment.

elaiosome – an oil-rich body on seeds or fruits that attracts ants.

elliptical – egg-shaped.

epidermal – of the epidermis.

epidermis (of a plant) – the outer single-layered group of cells covering a plant.

epidermis (of an insect) – the outmost layer of cells of the body of an insect; it is normally only one cell thick and covered by an impermeable cuticle.

eusociality – the highest level of social organisation; a society that usually has more or less sterile workers or soldiers that look after the reproductive members of the society.

evolution – the process by which different kinds of living organisms are thought to have developed and diversified from earlier forms during the history of the earth.

fauna (referring to insects) – a group of insects of a particular region or period.

founders – those who establish something.

frass – the mixture of faecal material and plant matter left by a plant-eating insect after it has eaten.

genus (plural genera) – one (or more than one) of the taxonomic groups into which a family is divided and which contains one or more species.

haemolymph – the circulatory fluid in many invertebrates that is functionally similar to the blood and lymph of vertebrates.

honeydew – a sugary substance excreted by some insects and exuded by certain plants.

host – an organism (plant, animal, bird or insect) that is fed on by a parasitic organism.

hyperparasitoid – a parasitoid species that attacks another parasitoid species.

hyperplasia – an abnormal increase in the number of cells of an organ, or tissue, causing it to increase in size.

hypertrophy – the enlargement of an organ, or tissue, from an increase in the size of its cells.

inflorescence – a flower cluster.

inquiline – an organism that lives inside another species' home, obtaining shelter and in some instances taking some of the other species' food, without actively destroying the other species.

instar – the stage in the development of an insect between any two moults.

kleptoparasite – parasitism by theft.

larva (plural larvae) – the immature form of an insect.

leguminous – belonging to the Fabaceae (formerly Leguminosae), a family of flowering plants having pods (or legumes) as fruits and root nodules enabling storage of nitrogen-rich material; examples are peas and beans.

lerp – a structure composed of crystallised honeydew produced by larvae of psyllid insects as a protective cover.

monoecious – having male and female reproductive organs borne on a single plant.

morphology – the form and structure of an organism or one of its parts.

moult (of insects) – shedding of the cuticle (noun) or to shed the cuticle (verb).

multivoltine – having two or more broods of offspring per year.

mutation – a change in the structure of the genes or chromosomes of an organism; genes are the units of heredity and are usually composed of sequences of deoxyribonucleic acid (DNA) that occupy a specific location on a chromosome; chromosomes are threadlike strands of DNA that carry the genes.

mutualism – occurs when two organisms interact so that each derives a fitness benefit from the interaction.

neotenic – retaining juvenile characteristics when adult.

nutrient sink – an entity that absorbs nutrients from a system.

nymph – the immature form of an insects that undergoes incomplete metamorphosis.

ostiole – an opening.

oviposit – to lay an egg.

ovipositor – the organ with which many female insects deposit their eggs.

parasite – an organism that lives on, or in, a host organism.

parasitoid – an organism that lives on, or in, a host organism and ultimately kills the host.

petiole – the stalk by which a leaf is attached to a stem; also called the leafstalk.

pheromones – chemicals secreted by an animal, especially an insect, that influence the behaviour or development of others of the same species, often functioning as an attractant of the opposite sex.

phloem – a tissue in a vascular plant that functions primarily in transporting organic food materials (e.g. sucrose) from the photosynthetic organ (leaf) to all parts of the plant.

phoresy – an association in which one animal clings to another to ensure movement from place to place, as some mites use some insects.

phyllode – a modified petiole that functions as a leaf.

phylogeny – the sequence of events involved in the evolutionary development of a species or taxonomic group of organisms.

physiology – the study of how organisms function.

physogastric – having a swollen abdomen.

pollen – a fine powdery substance consisting of microscopic grains that contain the male sexual cells of a plant.

pollination – the transfer of pollen grains to the female reproductive structure of a plant.

predators – organisms that live by preying on other organisms for food.

ptilinum – an expandable pouch on the head, above the base of the antennae, in some adult flies. It is normally used to break off the end of the pupal case so that the adult fly can emerge.

pupa – the non-feeding stage between the larva and adult in certain insects, during which the larva typically undergoes complete transformation within a protective cocoon or hardened case.

pupate – become a pupa.

sclerotised (of an insect's cuticle) – hardened and darkened.

sexual dichronism – producing offspring of one sex before producing the offspring of the other sex.

sexual dimorphism – males and females of the same species differing in appearance.

species – one of the basic units of biological classification.

stamen – the male reproductive organ of a flower consisting typically of a stalk (filament) and a pollen-bearing portion (anther).

stigma – an area at the top of a flower style for the reception of pollen.

stylets – needle-like projections of the mouthparts of some insects that are used to penetrate plant or animal tissue.

style – the tube leading into the ovary of a flower.

syconium (plural **syconia**) – the fleshy fruit of the fig, consisting of a greatly enlarged receptacle completely surrounding the inflorescence.

tannins – astringent, bitter plant compounds that bind or shrink proteins and various other organic compounds.

taxonomic – concerning the classification of organisms in an ordered system that indicates natural relationships.

thorax (of insect) – the second or middle region of the body, between the head and the abdomen; in insects it bears the true legs and wings.

univoltine – having one brood of offspring per year.

vascular plant – a plant that possesses a well-developed system of conducting tissue to transport water, mineral salts and sugars.

Index

www.ingramcontent.com/pod-product-compliance
Lightning Source LLC
Chambersburg PA
CBHW040953170526
45159CB00014B/3124